7101

The Computer Prophets

Books Written or Edited by
JERRY M. ROSENBERG

The Computer Prophets
The Death of Privacy
New Conceptions of Vocational and Technical Education
Automation, Manpower and Education

Jerry M. Rosenberg, Ph.D.

The
Computer
Prophets

The Macmillan Company
Collier-Macmillan Ltd., London

ACKNOWLEDGMENTS

Little, Brown and Company for permission to quote from *The Lengthening Shadow* by Marva and Thomas Belden.

McGraw-Hill Book Company for permission to quote from *A Business and Its Beliefs* by Thomas J. Watson, Jr.

W. Heffer and Sons, Ltd., and Sara Turing for permission to quote from *Alan Turing* by Sara Turing.

Library of Congress Catalog Card Number: 69-16493

First Printing

The Macmillan Company
Collier-Macmillan Canada Ltd., Toronto, Ontario

Printed in the United States of America

To my mother, Esther,
and the memory of my father, Frank

Contents

Introduction

When man decided that his ten fingers were insufficient for computing large quantities of data, he was motivated to invent various tools to assist him in his endeavors. Thus began a love affair between man and the magic of the number which continues today.

The passage of time has produced discoveries ranging from the abacus of antiquity to the modern high-speed calculator, with each successful new device slightly more efficient than its predecessor. Man's present adventure in the handling of numbers is the computer. No approach or machine of the past can compare with its known or projected capability, and few contributions in technology can match its impact in every phase of modern life.

Even though we think of the computer as an invention of the mid-twentieth century, even though we consider it still in its infancy, nevertheless the full story of its evolution is by no means that simple.

No one man can be given single credit for the genius of the present-day computer—unlike other inventions in history. A steep ladder had to be ascended. The creative teams of today who devote their fullest energies to the discovery of new theories, approaches, programs, units, etc., realize that the steps never end—that change is an essential requirement in this age of the computer. There was, however, a beginning, a place in time when a handful of dedicated men saw the need for machines to compute efficiently and devised the means for fulfilling their dreams.

The story of these computer prophets is worth telling. Some of them are legendary figures; others have been deprived of their true place in history. To appreciate the marvel of the computer is to understand the dynamics of the perplexed, questioning, involved and creative geniuses of the machine. In many ways they were like other mortals: they tried, and sometimes failed. They were the pioneers in the struggle to free man from the boring and often error-filled task of calculating by hand. These men were driven relentlessly to give a newer definition to man's role. By their achievement these men of systematic thought helped raise mankind to a higher level of increased knowledge and awareness.

Teacher and student alike should now be able to integrate the meaning of the present technology with that of the past. It is hoped that this book will stimulate many to explore the growing opportunities in this fast-developing industry. The scientist, technician and researcher may gain the necessary link between the challenges of the future and those already confronted by the men in this book. For the businessman it should be an aid toward a better appreciation and understanding of how he can more effectively utilize the modern computer.

The public also can profit from knowing more about both individual achievements and the history of an industry that has and will, even more in the future, alter its very way of life.

JERRY M. ROSENBERG

1

Blaise Pascal

The Genius Who Produced the First Efficient Calculating Machine

There is no name more famous in literature, science and religion than Blaise Pascal. He was certainly the greatest thinker of his day and one of the most productive men who has ever lived.

In his thirty-nine short years of life, Pascal was to master some of the most difficult concepts ever known to man. By the age of twelve he was already indoctrinated with the purposes of mathematics, and by sixteen he wrote a paper that was claimed as the beginning of modern projective geometry. Still in his teens, Pascal invented, built and sold the first calculating machine. He gave physics Pascal's Law, established the science of hydrodynamics and through experimentation demonstrated the existence of the vacuum. Interested in the game of chance, he was able to innovate the first mathematical theory of probability and introduce concepts vital to today's calculus.

By the time he was thirty-one, Blaise Pascal felt he had exhausted these interests and turned his back to science, devoting the remainder of his life to religious and theological pursuits. During these productive years Pascal was to establish himself as a brilliant and introspective author. His manner of writing in the simplest of forms was to influence and shape the French literary language for centuries.

The Europe that Blaise Pascal lived in was one of contradictions. The Thirty Years' War, which ended in 1648, left a permanent mark on the lives of all Europeans. It was a bloody and cruel war. Reprisals provoked more reprisals, humans were murdered and maimed with a ferocity to be found only in men who know that their turn may well be next. Peasant farmers stopped growing food when they realized that the only chance they had to obtain bread was to join one of the many marauding bands that swept Western Europe.

The war revealed a spectacle of futility and hopelessness. Hundreds of villages were completely destroyed, without a single inhabitant remaining. Towns were reduced to less than half their population.

The results of this struggle were severe and affected the course of the continent's development for centuries. In many respects it brought Europe back into the Middle Ages. The great decrease in population prevented the breaking up of large estates, affording feudal institutions a longer lease on life. Peasants remained serfs.

Following the Thirty Years' War, France returned to a period of calm. Her wealth grew, and the middle class acquired new comforts of life and found leisure for inquiring minds. Thus began a rebirth of intellectual curiosity in the arts, science, philosophy and religion.

Blaise's grandfather Martin Pascal was the treasurer of France. His father, Étienne, had achieved prominence as an important government official of Auvergne in central France, acquiring the position of Second President of the Court of Aides at Clermont, a small town of nine thousand inhabitants. He was recognized as a good mathematician because of his meditations on mechanics and on the theory of harmony. Blaise, born on June 19, 1623, was one of his three children to live beyond infancy.

The young Pascal was almost lost to history. At the age of one he became seriously ill with either tuberculosis or rickets. The doctors found that the sight of water made him hysterical, and he would often throw tantrums when he saw his father and mother together.

Medicine was still quite primitive, and witchcraft was looked upon as the doer of physical harm. Blaise's illness was diagnosed as a sorcerer's spell. To remove the curse it was necessary to find the witch, and the town turned against a poor old woman living on the charity of Mrs. Pascal and a frequent visitor to the Pascal house. This pathetic woman was under constant persecution. Although she believed herself innocent, to relieve this endless pressure, she threw herself before Étienne Pascal and promised to reveal all if he would not have her hung.

Her story was full of mysticism. She said that she had earlier asked Blaise's father, a lawyer, to defend her against an unjust suit, and, in retaliation for his denying her request she had put a death spell on the child. Although the charm could not be withdrawn, she said it could be transferred to bring about some other animal's death. A horse was first offered, but it was determined that a cat would be less costly. Descending the stairs with a cat in hand, she met two friars, who scared the old lady, and in her fright she threw the animal out of a window. Unlike the story of nine lives, this cat hit the pavement and died. Time was to be the healer, and Blaise did eventually recover.

Less fortunate was his mother, who died when Blaise was about four years old. Out of despair and grief, Étienne Pascal retreated from his work, discovering that his children's intelligence could absorb him more than his own worldly pursuits. In 1631 he sold his post to a brother, transferred most of his property into government bonds and moved to Paris.

There is little doubt that without his father's unorthodox style of teaching his children, Blaise would never have achieved the greatness that he was to find. In France at this time the educational system preached that a child's training should begin with the structure and comprehension of Latin grammar. Étienne felt so strongly that this was an unnecessary burden for so young a mind that he deferred it until Blaise was about twelve. In its place, Blaise was assigned the task of developing his mind in the areas of reason and judgment in order to appreciate that facts are revealed following the pursuit of a curious mind with a passion for knowing. Blaise was to learn the purpose of a fact and its value: curiosity would then do the rest. "His principal maxim was to keep the boy superior to his tasks," said Blaise's older sister Gilberte. She continued: "This general idea enlightened his mind and showed him the reason for grammatical rules, so that, when he came to learn the grammar, he knew why he was doing so, and applied himself precisely to those things which needed the most application."

Time was spent reading the Bible, ecclesiastical history and other writings of the Church. However, Blaise was most concerned with science. His father instructed him in the principles of the experimental method: how to observe, ways to classify and how to present generalizations from gathered facts. Gilberte recorded: "From his childhood, he could not surrender except to what seemed to him evidently true, so that, when he was not given good reasons, he sought them out himself, and when he had become interested in something he would not quit it until he had found a reason which could satisfy him."

When Blaise was eleven, he noticed that the striking of a hard object against a china dish presented a ringing hum which could be stopped by merely touching the plate with the hand. "Why?" asked the young Pascal. Dissatisfied with the answers he received, he started a series of experiments on the qualities of sound and wrote a paper explaining this phenomenon. Still a boy in short pants, he had learned the fundamental approaches of the scientific method. His sister said: "His mind was incapable of being satisfied, and it remained in a continuous agitation until it had discovered the true reasons."

This passion for learning drew Blaise toward mathematics and, in particular, toward geometry. His father, however, was not willing at this time to allow his son to become overwhelmed with geometry for fear that he would turn from learning Latin and Greek. Blaise insistently demanded, "Geometry! What is geometry?" but to no avail. Étienne was determined to keep the answer from his son. He locked up all his mathematics textbooks and cautioned his own friends not to mention mathematics in the presence of the youngster. All that Blaise could learn from his father was that geometry was the science of making true diagrams and of finding the proportions between them.

Undisturbed by his father's refusal to allow geometry to enter his life, Blaise started to apply the definition he was given. Drawing some circles and lines, he was enchanted with the way in which the forms evolved. He set out to invent his own words. He called straight lines "bars" and circles "rounds." He started writing down observations and discovered for himself basic truths, or axioms, of geometry.

He presented himself with problems and went about proving
them. Without the guidance of a teacher, or his father, he
was able through his understanding of logic and the scientific method to prove the thirty-second proposition of Euclid
—that the sum of the angles of a triangle is equal to two
right angles.

Overwhelmed, his father was convinced that his son was
a genius, and he immediately set out to unlock the secrets
of geometry to him. Blaise was to learn the classics under
formal instruction and master the sciences during his play
periods. Four years later, he was to compose a treatise on
how geometric cones were structured in which he uncovered
principles that were unknown in his day.

Blaise's modesty was revealed in his conclusion to this
six-page treatise:

We have several other problems and theorems, and several
consequences deducible from the preceding; but the mistrust
which I have of my slight experience and capacity does not
permit me to advance more till my present effort has passed
the examination of able men who may oblige me by looking at
it. Afterwards, if they think it has sufficient merit to be continued, we shall endeavor to push our studies as far as God
will give the power to conduct them.

Blaise's education was thoroughly unorthodox. He never
spent a day inside a school. He was never found playing
with boys of his own age, learning in the group process,
learning of the competition that develops from spending
time with fellow students. His boyhood was a period of
isolation. At the age of twelve he was appointed to a commission to judge procedures for reckoning longitudes. He
also uncovered an error in Déscartes' geometry.

At the age of thirteen he was introduced to the brilliant
intellectuals of France. His father, a member of the Academie Libre (later known as the Academie des Sciences), al-

lowed him to spend every Thursday with the famous doctors
of Paris, listening, absorbing and contemplating.

This exposure led Blaise to cherish physics and mathematics—especially geometry and in particular the shape of
cones—as the most central of man's activities. To uncover
their secrets he knew would be to follow the doctrines of
trial and error and reasoning.

The study of cones, or conics, did not begin with Pascal.
In the fourth century B.C, Plato and his students were
absorbed with determining essential curves in the world,
and Omar Khayyam, a court astronomer in the eleventh
century, had united conics with algebra.

About twenty-five years before Blaise was born, Kepler
discovered that planets moved in ellipses and employed
conics to explain partially the theory of universal gravita-
tion.

Girard Desargues, a friend of Étienne Pascal, was unwilling to accept the geometric interpretations of his contemporaries. He was to formulate the essential theory of
perspective and to be hailed some two hundred years later
as the father of projective geometry.

The boy Pascal was the only one in his day to accept the
incredible ideas and methods of Desargues. In the treatise
quoted above he expressed his appreciation: "I am willing
to admit that I owe the little I have found in this matter to
his writings, and that I have tried to imitate, as far as has
been possible to me, his method on this subject."

But Blaise was to bring order and interpretation to
Desargues' work in his theorem: If a hexagon be inscribed
in a conic section, the intersection of opposite sides are
three points in a line. "Mystic Hexagram" was the name
Pascal gave to his proposition.

This was his genius at work. His colleagues and men his
senior were impressed with his unusual curiosity and mental

energy. Blaise "wanted to know the reason for everything."

It took some time before the world was to accept this outstanding contribution. Although Desargues was pleased with Pascal's tribute to his theories, Déscartes, the great French philosopher-mathematician, was less praising: "I do not find it strange that there are some who demonstrate conics more easily than Apolonius, for he is extremely long and involved, and everything he demonstrated is easy enough in itself. But one can certainly propose other things concerning conics that a boy of sixteen would have trouble in solving."

Pascal put aside his work in conics and only occasionally thereafter did he return to geometry.

At this time, financial losses forced the Pascals to leave Paris. His father had invested his savings in bonds, and the government, requiring added funds, had lowered the value of the revenues. Étienne Pascal participated in public protests against the nonpayment of interest on government bonds. Cardinal Richelieu issued a *lettre de cachet* for his arrest which forced Étienne to go into hiding.

One day Blaise's sister Jacqueline appeared in a dramatic play before the Cardinal. After the performance, which was a tremendous success, full of tears she pleaded with the Cardinal for the pardon of her father. Her finest acting was to lead Cardinal Richelieu to say to Étienne: "I know all your merit. I restore you to your children and commend them to you." Étienne Pascal was appointed to the post of Royal Commissioner in High Normandy for the Tax Service. He arrived with his children in Rouen when Blaise was sixteen years old. Unlike the rest of his family, Blaise was neither happy nor unhappy with leaving Paris. It didn't matter to him, for he was about to invent a machine for performing arithmetical calculations.

The young Pascal had been preparing to write a concise

study of the entire field of mathematics when his attention was turned to assisting his father in his official duties at Rouen. Étienne constantly requested that his son assist him in the mounting financial problems that were part of his tax assessments. The enormity of the hand calculations bothered Blaise, who was able to explore easily the unknowns of abstract mathematics but was annoyed with the drudgery of making arithmetical totals. As one might suspect, he wasted little time in deciding that the solution to his problem lay in the construction of a machine.

Until Pascal's time, operating calculators were unheard of. However, he was not the first person to inquire into the mechanization of addition. The abacus, probably the most enduring counting system, seems to have originated in the Tigris-Euphrates Valley some five thousand years ago. A curious finding is that this highly efficient system was independently invented and developed by people in scattered areas of the world. In Peru, for example, the Indians tied knots in ropes to indicate their computations.

The abacus, as we know it today, was invented in China about 2600 B.C. (the Japanese equivalent was a soroban). The abacus has five beads on one end of one rod, then a division in the frame, then two beads on the other end of the rod. There are two fives in ten that make a positive count, and the carrying is identified by means of the other two.

The Greek abacus, which is still used today in many countries of Eastern Europe, has nine beads on one rod, then a division, then one bead. Although the Romans had strung pebbles together to form counting ranks, the abacus did not reach Central Europe until the beginning of the Christian era.

Nearly four thousand years passed before the next computational device was built. One reason was the difficulty in using Roman numerals as the principle means of adding on the European continent. For example, a machine could not easily add MCMXXI to MCXVII. When the Arabic numerical system started to find its way throughout Europe, this barrier was gradually eliminated.

Gerbert of Aurillac, later Pope Sylvester II from 999 to 1003, was the first person to try to mechanize the abacus. Gerbert was a shepherd boy. A monk noticed that he was quite ingenious and possessed an enormous passion for learning. Gerbert made an instrument for playing music by peeling bark off limbs of trees and fitting it with reeds; then he ran water through a pipe to force air down and vibrate the reeds.

Practically all the knowledge and science of the world was then possessed by the Moors who occupied southern Spain and Northern Africa. The Moors had two great universities, one at Cordova and the other at Seville, but no Christian could enter them, so Gerbert associated for some years with Moors, lived their life and adopted their customs along with the Mohammedan religion. Suddenly he disappeared from the monks and in Moorish garb applied at one of the universities. As he seemed to be a good Mohammedan, he was taken in and went through both universities. After he graduated, he returned to Christian Europe.

When Gerbert came back he brought with him what we refer to as Arabic numerals, as well as plans for a calculating machine that the Moors had been working at but had never succeeded in making work. Gerbert spent many years of his life trying to make the machine work accurately but was unsuccessful. He had one thousand counters, made of horn, arranged into twenty-seven divisions. Since the concept of the zero was hardly known, his instrument was not much of an improvement over the hand operation.

But another Spaniard took up the idea and made a calculating machine. His name was Magnus. His device, formed of brass, was in the shape of a human head, and the figures appeared where the teeth would be. However, there was no evidence to show how accurate Magnus' apparatus was.

The priests of the day (about A.D. 1000) thought that his device was something superhuman and they smashed it with a club.

The master genius of them all, Leonardo da Vinci, was never looked upon as a contributor to the world of calculation. With the 1967 rediscovery of two bound volumes of his notebook materials in Madrid in the National Library of Spain, it was proved once and for all that he did have some insights into the evolution of a mechanical adding machine during the fifteenth century.

His drawing described a device that would maintain a constant ratio of ten to one in each of its thirteen digit-registering wheels. For each complete revolution of the first handle, the unit wheel was to be turned slightly to register a new digit ranging from zero to 9. Consistent with the ten to one ratio, the tenth revolution of the first handle would cause the unit wheel to complete its first revolution and register zero, which in turn would drive the decimal wheel from zero to 1. Each additional wheel was marked with hundreds, thousands, etc. Weights were used to demonstrate the equability of the machine.

No working model was developed and the chances are that Pascal never saw da Vinci's sketch.

In 1623, Wilhelm Schickard, a professor of biblical languages and astronomy, designed a machine that could add, subtract, and, in fact, could multiply and divide. The model was destroyed in a fire and a new one was never rebult.

John Napier, a Scot, developed logarithms in 1617, a tabular system of numbers by which many calculations in arithmetic are simplified. He invented a device that was later called "Napier's Bones." It was a mechanical arrangement of bone strips on which numbers were stamped out. When placed into the proper combination these strips could perform direct multiplication. For example, to multiply 526 by 3, the five and two and six strips, identified by numbers at the top, were placed against each other. By reading the multiples in the third row from right to left, adjusting to add the ten digits to the adjacent unit digits on the left, the multiplication product could be read.

Blaise was to go beyond Napier and develop a calculator that is essentially like the one used today. He arranged the digits of a number in wheels. Each wheel would in turn make a complete revolution, thereby shifting its neighboring wheel one tenth of a revolution. This most elementary of processes was entirely new in the realm of applied arithmetic. It was the product of a genius' mind, and he called it his "pascaline."

Pascal never documented how his machine worked. He excused himself from this responsibility by noting that such a description would be useless without entering into a number of technical details that would be unintelligible to the general reader. He was more comfortable with suggesting that the best way to understand his work was to see it.

Blaise, in 1642, was only nineteen when the general idea for designing a calculator came to him. For the next three years nothing else mattered to him but how well his calculator was developing. In fact, he was to remain engrossed in his invention until he was nearly thirty, trying to design the parts to enable it to handle fractions and calculate square roots.

Model after model was constructed. Promise turned to failure, only to be challenged again with a new approach, and subsequently a new model. Assisted by mechanics, he made more than fifty models, some of wood, others of ivory, ebony and copper. Connecting rods and flat metal strips, plane and curved, chains, cones, concentric and eccentric wheels were put into place as directed by his drawings. He invented a "jumper" for making the machine record its results.

But Blaise was to find his work interrupted. It was discovered that a clever watchmaker in Rouen had stolen the principle for a calculator from him and had quickly built his own calculator and was passing it off as the first of its kind.

So distressing was this tinker's action that in a moment of utter disgust, Blaise fired all his workmen and ceased the construction of his calculator. His annoyance was not merely that his idea had been stolen but, more important to Blaise, that an inconsequential watchmaker would be compared to him. This "little abortion," he described it; this "little monster lacking its principal members, the others being formless and without proportion."

The senior Pascal was upset with his son's actions and tried, at first without success, to persuade him to return to his calculator and complete it. Étienne sent a model of the machine to his friend and supervisor Chancellor Seguier in Paris. After careful examination of the calculator, the Chancellor granted Blaise the exclusive right to manufacture his adding machine and ordered that he hurry to complete and perfect his work.

Pascal had received a special *privilège du roi,* (privilege of the king). The statement excluded all others from manufacturing a calculator. Competitors "of whatever quality and condition were prohibited from making it, or causing it to be made, or selling it."

At least ten of Pascal's original models are still known to exist. The calculator was a lightweight polished brass box, about 14 by 5 by 3 inches. As one looks down toward the top one sees a row of eight movable dials. The dial to the right contains twelve slots representing the French monetary *deniers*. At its left, another dial, with twenty slots, represents the *sous*. The remaining dials, all with ten slots, are for the franc, the *livres*. The calculator can be used for nearly any form of currency. To perform calculations with the American decimal system, the two dials on the right are not used, but the remaining six wheels are.

On the top of the lid over each wheel is a series of eight windows showing a number below it.

Anyone using a Pascal calculator is impressed with its simplicity. For example, in completing the following addition: 125 plus 311, all the dials are turned so that their zeros appear in each of the windows. Since we are using the decimal system, we start with the third dial from the right. The first quantity to enter is 125. Starting with the third dial, for units, a pointer is inserted into the slot marked 5. The wheel is then rotated clockwise until the pointer is stopped by a bar, as is the case when dialing a telephone number. The next wheel on the left, for the tens, is revolved from 2 until it stops and then the next left wheel, for the hundreds, from 1 until it stops.

At this time, the numbers 125 appear in the windows reading from left to right. Should 311 be added and the same process carried out, the windows would read the total 436.

In order to subtract, a flat metal ruler just back of the windows is pulled forward. By doing this, a second set of windows appear which are extensions of the first set. Suppose you want to subtract the number 1 from 3, the dials are turned until 3 appears in the window. After dialing 1, the remainder 2 automatically appears on the indicator.

To multiply with Pascal's calculator, additions are made in a slightly different way. To multiply 2468 by 678, the number 2468 is entered eight times, commencing with the digit dial on the right. Then 2468 is entered seven times beginning with the hundreds. Division is carried out in reverse until the indicator returns to zero.

The genius of Pascal is observed upon opening the box. The gears of each wheel are like teeth and fit securely between the teeth of another wheel. Pascal had to have each gear filed down to size, inserted into its hole drilled in the wheel, and laboriously soldered into place.

The most difficult mechanism is the ratchet device that communicates by one revolution of a wheel a movement of one digit to the wheel of the next highest order. As the lower wheel passes from 9 through 0, a ratchet is forced to forward the higher numeral wheel the proper distance. His sister said: "The direct actuation of a numbered wheel through its various degrees of rotation and the secondary feature of effecting a one-step movement to the numbered wheel of higher order is the foundation on which nearly all the calculating machines have since been constructed."

Pascal had learned to use and understand the three dimensional properties of projective geometry. From his knowledge of how planes and curves behave he was able to design wheels and gears and apply them in the construction of a calculator. He said: "To make the movement of my machine's operation the more simple, it had to be composed of a movement the more complex."

The world of intellectuals was astounded by this achievement. Many of the scientists of the day had drawn on paper inventions for airplanes, submarines and other futuristic devices, but none had proceeded to realize his idea with a working model. Pascal did and he was acclaimed throughout Europe. Gilberte stated: "Pascal knew how to animate copper and give wit to brass."

Pascal and his father saw this machine as a certain avenue to wealth. They had spent great sums of money in building the many models needed before it could be perfected, and Blaise was certain that within a few short years his invention would be selling the world over. He knew that financial success would depend on the testimonials of the powerful in France. Models were sent to businessmen, royalty and people influential in the world of finance. His letters illustrated how easy the calculator was to operate, how fast it would make its additions, how error-free it would be and how indestructible it was. In Paris, Blaise gave lecture-demonstrations of his calculator in the salon of the niece of Cardinal Richelieu. His brochure of advertisement was addressed to "Friend Reader," explaining the drudgeries in hand multiplication and the savings of time if his machine were bought.

To the amazement and disappointment of the Pascal family, the calculator failed to impress the business world and was a financial failure. Potential buyers thought it was too complicated and could be repaired only by Pascal. Even if the machine could perform the work of six men, many claimed that the half-dozen men were still less expensive than the calculator. And there was the fear of some that the machine would lead to unemployment of bookkeepers and other clerks. Ironically, versions of Pascal's machine are sold today as inexpensive pocket calculators.

Although this invention did not bring Blaise Pascal wealth, it certainly brought him fame. His reputation passed beyond the circles of intellectuals and reached out to the world. In February, 1644, Bourdelot, a medical doctor, urged Blaise to demonstrate his machine to the Prince de Condé while in Paris. His device was praised in prose and verse by Mersenne and Huygens. A hundred years later, Diderot still felt it necessary to include a detailed sketch of the machine among the plates of his encyclopedia.

The calculator remains as Blaise Pascal's only useful mechanical invention. His other devices were hardly original with him. He built a new type of wristwatch, a spring-movement for clocks and a windlass. However, none of these inventions contained any new mechanical principle. There are some who claim that Blaise also invented the wheelbarrow, but recent evidence has shown that the wheelbarrow was in use during the Middle Ages.

Having concluded his work on the calculating machine, Pascal turned his interests to physics. For centuries the intellectual world had argued that a vacuum could not exist. Aristotle claimed that nature would not tolerate a vacuum: *Natura abhorret vacuum.* Descartes said that, "if everything should be removed from a vessel, the sides must immediately touch, for a vessel cannot be filled with Nothing; that is a logical impossibility, ergo false." The world would sooner fall apart than allow Nothing to exist. The scientific community referred to the vacuum as "the horror of the void."

To account for the existence of a vacuum above the water, Pascal presented the supposition:

That it contained no portion of either of these fluids, or of any matter appreciable by the senses; that all bodies have a repugnance to separate from a state of continuity, and admit a vacuum between them; that this repugnance is not greater for a large vacuum than a small one; that its measure is a column of water about 32 feet in height, and that beyond this limit a great or small vacuum is formed above the water with the same facility, provided that no foreign obstacle interfere to prevent it.

Pascal, now twenty-four, held a demonstration before an audience of five hundred in Rouen. He was to prove beyond question that the Italian Torricelli's claim of the existence of a vacuum could be experimentally shown. In his experiment Blaise took two tubes 46 feet long and had them mounted on pivots at their midsection. One tube, filled with water, was turned upside down so that the open end was submerged in a container of water, and the stopper was removed. Immediately, the water level fell to a height of 34 feet above the water level in the container.

Turning to his friends in the audience, he asked, "What would happen if wine should be used in the test?" Several familiar with the properties of wine said that since wine was more active than water, more vapors would be released and therefore the liquid would descend farther in the tube. Pascal then filled one tube with water and the other with red wine; the wine stood higher in its tube than the water—a direct disagreement with members of the audience.

He presented a siphon containing one arm 50 feet high, another 45 feet. The water could not be siphoned because the connecting arm was more than 34 feet above the water. When he lowered the arm to 34 feet above water level, the water started running through the siphon and a vacuum was created. Pascal was eventually to demonstrate that air had weight and that its pressure could create a real vacuum: "Liquids weigh according to their height."

Pascal had proved beyond the possibility of a doubt that air pressure supports the mercury in a barometer and lifts the water in a pump. He placed two mercury barometers exactly alike at the foot of a mountain. The mercury stood at the same height in each. Then one barometer was left at the foot of the mountain and the other was carried to the summit, about three thousand feet high. The mercury in the second barometer then stood more than three inches lower than at first. Carrying the barometer down the mountain, he noticed that the mercury slowly rose until his arrival at the foot, when it stood at the same height as at first. His researchers stopped about halfway down the mountain, allowing the barometer to rest there for some time and, observing it carefully, found that the mercury stood about an inch and a half higher than at the foot of the mountain. Throughout this experiment the height of the mercury in the barometer which had been left at the base of the mountain did not change.

Pascal also invented the hydraulic press, a machine with which he said he could multiply pressure to any extent. He could so arrange his machine that a man pressing with a force of a hundred pounds on the handle could produce a pressure of many tons. Pascal experimentally proved and stated in a clear formula the phenomena behind the hydraulic press. "If a vessel full of water, entirely closed, has two openings, one a hundred times the area of the other, and if one puts in each a tight-fitting piston, a man pushing the small piston, will equal the force of a hundred men. . . . Whence it appears that a vessel full of water is a new principle of mechanics and a new machine to multiply force to any desired degree."

Pascal's findings in science, though important, were of lesser value than his style of inquiry. His power of concentration and his understanding of the scientific method are his most important permanent monuments. In the seventeenth century there was no accepted scientific methodology. Interpretation was based on observation and an understanding of theological authority. Pascal was to bring the experimental method a long way in discovering nature's secrets. "The secrets of nature are hidden; although she acts forever, we do not always discover her effects. Time reveals them from age to age, and although she is always equal within herself, she is not always equally known. The experiments which give us our knowledge of her multiply continually; and as these are the only principles of physics, the consequences multiply in proportion."

He summarized his belief in the importance of sound experimentation. "The discovery is made by virtue of a precise reasoning, established on an idea born from observation and controlled by experiment." But Pascal felt there was something missing in his scientific method, something he called "the spirit of finesse," or "heart." He had the ability to put order into disorganized facts, to accept axioms and principles. This theory of intuition, as we call it today, was one of Pascal's most important contributions.

Pascal's experimentation with physics lasted two years and was abandoned so he could return to the pursuit of his soul's salvation. Blaise Pascal's earlier adoration for science had turned into dissatisfaction. "We make an idol of truth itself; for truth outside of charity is not God; it is his image, and an idol which we must not love nor adore; and still less must we love or adore its contrary, which is falsehood."

Pascal cannot rank with Newton or Galileo in physics, because he was unwilling to spend a lifetime at these pursuits. He had experimentally proved the vacuum through new concepts of air pressure. Applications of his experimentation were later used in designing barometers and hydraulic presses. He gave us the principles to explain the elasticity of gases and how pressures act on liquids—and some of the best examples of the scientific method.

Undoubtedly, Pascal's great contribution outside the areas of religious philosophy and literature was his joint creation with Pierre Fermat of the mathematical theory of probabilities. In his theory of probability he stated clearly and solved convincingly a real problem—how to bring the superficial lawlessness of pure chance under the control of order and regularity. This contribution alone stands at the very base of all experimentation in the sciences.

Pascal became interested in this area when the Chevalier de Méré, a part-time professional gambler, presented a problem to the mathematician. For example, each of two dice players needs a certain number of points to win a game. Should they leave the game before it is officially over, how would the stakes be divided between them? After each player is given a score when he quits the game, the question is to determine the probability that each player has at a given stage of winning the game.

This Pascal figured out, and today the application of probabilities has become the basis in insurance, statistics, educational analysis and a considerable part of modern theoretical physics and chemistry.

Having accomplished so much in so short a time, he completely rejected science and instructed his family to look at his scientific duties as "the games and diversions of his youth." His choice was a disappointment to many. To them Pascal would say that "all sciences would not console him in the time of affliction; but that the science of the truths of Christianity would console him at all times, both for affliction, and for the ignorance of the sciences."

In science, Pascal had found the vacuum and nature's void. Now Pascal began to try to fill the greater void within his mind. At the age of thirty, Pascal left his world of science. He was seized by "a great scorn of the world and an unbearable disgust for all people who are in it." He sought isolation and escape. He felt that he had previously known nothing but shadows. He was determined to devote himself to God and feared that even his mathematical pursuits were selfish.

One night in 1654, Pascal felt that God came to his room and spoke to him for two hours. Pascal's inner conflicts were great, and in this mystical experience, he felt that he had at last been freed from the bondage of corruption. He was disenchanted with what life had brought him. His calculator was a financial failure. Although he was admired by the scientific community, he felt that they were unable to comprehend the significance of his mathematical inquiries. His feelings were presented in one of his writings: "For those who are capable of inventing are rare; the greater number wish only to follow, and refuse glory to those inventors who seek it by their inventions; and if they persist in trying to obtain glory, and in scorning those who do not invent, the others give them funny names, and would be glad to beat them. So let no one pride himself on his ingenuity, or let him content himself within himself."

One of the outstanding religious reformers of that period was Cornelius Jansen, a Dutchman who became Bishop of Ypres. The cardinal point of his dogma was the necessity for "conversion." Salvation, on the other hand, was one of Jansen's lesser interests. He was convinced that God had chosen him to blast the Jesuits in this life and prepare them for eternal damnation in the next. His principle goal was the continuous hatred of all those who disregarded his dogmatic bigotries. Of all the religious alternatives of that period, unfortunately this is the one that Blaise Pascal chose.

On the day of his conversion, November 23, 1654, he was driving a four-in-hand when his horses bolted. The two forward animals plunged over the parapet of the bridge at Neuilly, but the traces broke and Pascal was thrown onto the roadway. Pascal's mystical temperament convinced him that this miraculous escape from death was a direct warning from Heaven, to pull himself away from his internal questioning and devote his years to God.

Blaise Pascal was disillusioned and filled with contempt, both for the outside world and for himself. He spoke of his shortcomings and sins: "If one does not know himself to be full of pride, ambition, concupiscence, weakness, pettiness, injustice, one is very blind. And if, knowing this, a man does not desire to be delivered, what can one say to him?"

He turned to religion and sought out God. But his new path was to lead him to further unhappiness. He could find no Lord and in fear he said: "The eternal silence of these infinite spaces terrifies me."

Pascal was a believer in God but was suffering from a sense that God had abandoned him and that he was merely spending his life wandering in a desert of emptiness. His sister Jacqueline describes Blaise after a visit to her convent:

On this visit he opened himself to me in a pitiful way, admitting to me that in the midst of his occupations, which were great, and among all the things which should contribute to make him love the world, and to which he was outwardly much attached, he was so impelled to quit all that, both by an extreme aversion from the follies and beguilements of the world and by the continual reproach of his conscience, that he found himself detached from all things in a way that he had never been before, nor anything like it; but that also he was in such a great abandonment on God's part that he felt no attraction in that direction; that nevertheless he urged himself with all his power, but that he felt very much that it was his reason and his own mind that were exciting him to what he knew to be the best thing, rather than the movement of God's spirit, and that in the detachment from all things in which he found himself, if he had the same feelings for God as previously, he thought he could have undertaken anything, and that he must have been bound by horrible attachments in those times to resist the grace which God gave him and the impulses he sent.

A few days after Pascal's death a servant was organizing his clothes and felt a piece of paper under the lining of his coat. This sheet of paper described his two hours in God's presence, his mystical illumination. Pascal had been carrying this note in his coat for eight years since his bridge accident in Neuilly.

This scribbled piece of paper remains one of the treasures of the National Bibliothèque in Paris. Under a cross, which appears at the top of the paper, Pascal inscribed:

The year of grace 1654.

Monday, 23d November, day of Saint Clement, pope and martyr, and others in the martyrology.

Vigil of Saint Crysogone, martyr and others.

From about half past ten o'clock in the evening till about half past twelve.

FIRE

God of Abraham, God of Isaac, God of Jacob, not of the philosophers and of savants.

Certitude. Certitude. Sentiment. Joy. Peace.

God of Jesus Christ.

My God and your God.

Thy God will be my God.

Oblivion of the world and of all save God.

He is found only by the ways taught in the Gospel. Grandeur of the human soul.

Just Father, thy world hath not known Thee, but I have known Thee.

Joy, joy, joy, tears of joy.

I am separated from Him.

They have forsaken Me, the fountain of living water.

My God, will you forsake me?

Oh, may I not be separated from Him eternally.

This is life eternal, that they know Thee the only true God, and Him whom Thou hast sent, Jesus Christ.

Jesus Christ.

Jesus Christ.

I am separated from Him; I have fled, renounced, crucified Him.

Oh that I may never be separated from Him.

He is only held fast by the ways taught in the Gospel.

Renunciation, total and sweet.

Total submission to Jesus Christ and to my director.

Eternally in joy for a day's trial on earth.

Non obliviscar sermones tuos. Amen.

The fire passes and Pascal is alone. His body is cold and momentarily paralyzed. He turns and looks at his clock. It is half-past twelve.

Pascal had renounced the world. It was more than the logical result of his vision. It was for him a new kind of logic that had just been found. He had been searching for a way to reconcile the differences of the world. His failure to reconcile the happiness and misery of the human spirit was his undoing.

God existed for Pascal and committed himself to making peace with Him. "We are lepers; how can we think of anything but our leprosy? What would you say of a leper who, indifferent to his gnawing disease, would talk of botany or astronomy to his doctor?"

In his *Apology for Christianity,* Pascal said: "Men despise religion, they hate it, and are afraid it is true." He noted man's indifference: "This negligence . . . irritates me more than it moves me to pity; it amazes me and appals me, it is a monster to me." The emotions that rings through the *Apology* are unmatched in today's religious literature. These pages of passion remain to this day a salute to Pascal, for our mind and soul continue to search for the answers that Blaise Pascal was unable to find.

Pascal was to make one last contribution to mathematics. Among the many maladies that afflicted him were insomnia and bad teeth. Tossing and turning one evening in bed with the misery of a toothache, Pascal began to think passionately about the geometrical curve—the cycloid—to take his mind from the intensive pain. To his great bewilderment the pain suddenly stopped. Pascal interpreted this as a signal from Heaven that he was not sinning in thinking about geometry rather than the salvation of his being. For a period of eight days he once again delved into the world of mathematics and succeeded in solving many of the main problems dealing with the geometry of the cycloid.

Because Pascal was still in doubt about the correctness of his behavior, many of these answers were released under the pseudonym of Amos Dettonville.

By 1658 Pascal's health was worsening. He confided to his sister that "from the age of eighteen he had not passed a day without suffering." In June, 1659, a friend described him as existing in a state of languor, tortured by headaches and vital pains, and living on ass's milk and bouillon."

He felt he could no longer trust his memory. "Escapes thought, I wished to write it; I write, instead, that it escaped me." He wrote on any bit of paper available. When he was unable to write he dictated his phrases, his *Pensees,* to his nephew or to a servant.

Shortly before death Pascal revealed his latter thoughts about mechanics and mathematics:

> To speak frankly about geometry, I find it the highest exercise of the mind, but at the same time I recognize it as so useless, that I make little difference between a man who is only a geometer and a skillful artisan. Thus I call it the finest trade in the world, but after all it is only a trade; and I have often said it is a good thing for the trial of our strength, but not for the use of it; so that I would not take two steps for geometry's sake.

In the summer of 1662, Pascal had invited a poor family to live with him. One of the children came down with smallpox, and it was suggested that the child be taken from the home. In the spirit of charity, Pascal insisted that he would be better able to bear the pain more than the child and allowed himself to be transported to a home a half-mile away.

He wanted no sympathy or pity. He told his sister Gilberte: "Do not be sorry for me. Sickness is the natural state of Christians, because therein one is as one should always be, that is, in suffering, illness, the deprivation of all good things and all pleasures of the senses, exempt from all passions, without ambition, without avarice and in the contiual expectation of death. It is not so Christians should pass their lives?"

In his last days he fought for life. He wanted more time to help the poor. While completing his will he asked: "How does it happen that I have not yet done anything for the poor, although I have always had so much love for them?" When he was told it was because he didn't have enough money, he said, "Then I ought to give them my time and my trouble; that is where I have failed. And if the doctors tell the truth, and God permits me to recover from this illness, I am resolved to have no other occupation and employment the rest of my days but the service of the poor."

But he was not to be granted his last wish. He died, on August 19, 1662, of a brain hemorrhage two months after his thirty-ninth birthday. His last words: "May God never abandon me!"

Blaise Pascal will always be remembered as the perfection of a Frenchman. He attained a genius, an abstract intellect with practicality, a logic with mysticism, the understanding of geometry and probability, poetic imagination and an awareness of reality—as measured by his calculator. In Pascal, France had produced perhaps her greatest mind.

His brain was expertly efficient. He was a genius of the

first order, but he was not a saint of the church. Pascal knew the difference himself:

Great geniuses have their empire, their glory, their grandeur, their victory, their luster, and they have no need of carnal powers, with which they are incommensurable. They are seen not with the eyes, but with the mind, and that is enough.

The saints have their empire, their glory, their victory, their luster, and they have no need for carnal or mental powers, with which they are incommensurable, for they neither add to them nor subtract from them. They are seen by God and the angels, and not by the bodies nor by curious minds. God is sufficient for them.

In the history of science Pascal will stand first in the second rank. To the mechanically oriented Blaise Pascal will be remembered as the person who invented the first truly operable calculating machine. But in the end Pascal cared little for this innovation. If ever a gifted man buried his talents, Pascal did; but perhaps he possessed a wisdom that transcended the material. Pascal wanted to be seen by the Gods and angels, not by men. He would not have cared how we looked upon his calculating machine, but one cannot fail to ask how happy he would be today had he known what wonders his mechanical ideas would bring more than three hundred years after he affixed his ratchet to a series of rotating wheels.

2

Gottfried Wilhelm Leibniz

He Believed His Mind to Be a Small Mirror of the Divine Mind

Few periods in the history of Germany could have been less favorable to the pursuit of creative work than the middle of the seventeenth century. The Thirty Years' War had a disasterous effect on the country, characterized by considerable poverty among its people, declining respect for all forms of education and the growing struggle of all religious groups.

In 1519, Frederick the Wise of Saxony refused the German crown when it was offered to him on the death of Maximilian I. By doing so he doomed Germany to an age of prolonged strife and warfare which, in the course of 130 years, depleted her material wealth, dismembered her empire, depopulated the country, brutalized her people and ended only through the absolute exhaustion of the nation. The Westphalian Treaty, which concluded the Thirty Years' War in 1648, left the country without a giant philosopher when she was most in want of one.

During this turmoil, Gottfried Wilhelm Leibniz was born on June 21, 1646, in Leipzig. He would be the first to turn the thoughts of his countrymen into channels out of which the culture of Germany was to spring. As diplomat, philosopher, mathematician, mechanic and historian he contributed to society.

His heritage of service was considerable: his family had been part of the Saxony government for three generations. His father, a professor of moral philosophy at the University of Leipzig and a jurist, died when Gottfried was six years old, leaving his only child in the hands of his widow, only thirty-one years of age. From his father, young Gottfried had acquired a passion for history.

Since the Leibnizes lived in a university community, and were part of the academic circle, Gottfried was exposed to a scholarly environment early in life. His father's library was available to him, and before the age of ten he had consumed books on Cicero, Pliny, Herodotus, Xenophon, Plato, the Romans and the Greeks.

Years later he was to admit that the ancient writers had a great effect on his understanding of the world's knowledge. He had learned to use words to attain clearness and to use them properly. Early in life he established for himself the formal rule of definiteness and clarity of diction, and the practical one of doing and saying everything for a purpose and an end. This was to lead him to the study of logic—one of his passions throughout his life. He learned how to use knowledge efficiently by classifying and systematizing it, by using signs and characters in place of words, by generalizing terms and by bringing every inquiry under a method and principle. His talents were therefore most naturally to turn to mathematics and other related applications.

By the time he was eight Leibniz began the study of Latin, and before he was thirteen he had become sufficiently proficient that he was able to write Latin verse. After Latin he approached Greek with the same vigor of mind and without the assistance of a teacher. There was no school in his day in which a boy of his genius could be educated; he was therefore left to the guidance of his own interests and he became self-taught. He later said, "More frequently will he who does not understand an art find something new than he who does, he who is self-taught sooner than those who are not—for he enters a road and gate unknown to others, and gains a different view of things. He admires that which is new, while others pass it by as something well known."

By the time he was ready to enter the University of Leipzig at fifteen he was no longer satisfied with classical studies. He had begun to seek outlets in logic and matriculated as a student of law. But finding that the instruction he received was thoroughly inadequate, he received private study to stimulate and satisfy his inquiring mind.

During his first two years, his law studies did not demand all his energies. He read philosophy and became aware, for the first time, of the findings of the natural philosophers, Kepler, Déscartes, Lull and Galileo.

In his third year he wrote a brilliant dissertation, "De principio individui," on the principle of individuation. He wrestled with the problem, Do individual things exist, or only their attributes or qualities? Leibniz claimed that only individuals exist. The most important aspect of his dissertation is that it was an expression of his philosophical opinion and his extensive knowledge of scholastic learning.

Leibniz realized that the new philosophies could be fully understood only by learning more of mathematics, and he spent a summer at the University of Jena to attend the classes of Professor Erhard Weigel, a man of local prominence but with limited national reputation in mathematics.

In 1666, at the age of twenty, Leibniz was preparing to finish his Doctorate of Law but was refused the degree. Although the official reason given was his youth, the truth was that he knew more about the law than the faculty and they were jealous of him.

Thoroughly disgusted with this petty behavior, Leibniz left his home in Leipzig and went to Nuremberg, where that same year he not only earned his degree but was begged to accept the university professorship of law. But he declined, pleading different plans.

One of Leibniz' remarkable traits was his ability to work anywhere, under the worst conditions, and still be able to discipline himself as if his were a private world. His dissertation on a new method of teaching law had been written on his brief trip from Leipzig to Nuremberg.

His essay, which he called a "schoolboy's essay," was on "De arte combinatoria." Leibniz tried to develop

a general method in which all truths of the reason would be reduced to a kind of calculation. At the same time this would be a sort of universal language or script, but infinitely different from all those projected hitherto; for the symbols and even the words in it would direct the reason; and errors, except those of fact, would be mistakes in calculation. It would be very difficult to form or invent this language or characteristic, but very easy to understand it without any dictionaries.

He was later to say: "If controversies were to arise, there would be no more need of disputation between two philosophers than between two accountants. For it would suffice to take their pencils in their hands, to sit down to their slates, and to say to each other: Let us calculate."

The paper was a defense of the method developed by Ramon Lull, who had been born in 1232 near the city of Palma, on the island of Majorca. The story is told that this Spanish theologian and visionary climbed Mount Randa in 1274 in search of spiritual understanding. After a long siege of meditation and fasting, he experienced a divine illumination in which God revealed the Great Art by which Lull might confound infidels and determine with accuracy the dogmas of his faith. According to the legend, the leaves of a small lenticus bush became spectacularly embossed with letters from various languages. Ultimately these were to form the languages in which Lull's Great Art was to be taught.

Shortly after his illumination Lull completed the first of forty treatises on the working and application of his method —his *Ars magna*. This was the earliest attempt in the evolution of logic to use geometrical diagrams for the purpose of discovering mathematical truths and the first attempt to use a mechanical tool—a type of primitive logic-machine—to carry out the operation of a logic system.

Lull has been credited by many as being the innovator of modern symbolic logic. Others have claimed him to be no more than a gifted crank. While still in his teens he was a page in the service of King James I and eventually became an influencial member of the court. He married young and fathered two children—which did not stop Ramon from his courtier's adventures. At the age of forty he said: "The beauty of women has been a plague and tribulation to my eyes, for because of the beauty of women have I been forgetful of Thy great goodness and the beauty of Thy works."

He had an especially strong passion for married women. One day as he was riding his horse down the center of town he saw a familiar woman entering church for a High Mass. Undisturbed by this circumstance, he galloped his horse into the cathedral and was quickly thrown out by the congregants. The lady was so disturbed by this scene that she prepared a plan to end Lull's pursuit once and for all. She invited him to her boudoir, displayed the bosom that he had been praising in poems written for her and showed him a cancerous breast. "See, Ramon," she said, "the foulness of this body that has won thy affection! How much better hadst thou done to have set thy love on Jesus Christ, of Whom thou mayest have a prize that is eternal!"

In shame Lull withdrew from court life. On four different occasions a vision of Christ hanging on the Cross came to him, and in penitence Lull became a dedicated Christian. His conversion was followed by a pathetic impulse to try to convert the entire Moslem world to Christianity. This obsession dominated the remainder of his life. His *Book of Contemplation,* divided into five books in honor of the five wounds of Christ, dealt with a number symbolism. It contained forty subdivisions—for the forty days that Christ spent in the wilderness; 366 chapters—one to be read each day and the last chapter to be read only in a leap year. Each of the chapters had ten paragraphs to commemorate the ten commandments; each paragraph had three parts to signify the trinity—for a total of thirty parts a chapter, signifying the thirty pieces of silver.

No doubt Lull was bordering on hysteria—perhaps on madness. In the final chapter of his book he tried to prove to infidels that Christianity was the only true faith. In explaining how his book was to be used, Lull said it should be "disseminated throughout the world," and he guaranteed the reader that he had "neither place nor time sufficient to recount all the ways wherein this book is good and great."

What became of interest to Leibniz was Lull's method of logic. Lull believed that in every branch of knowledge there were a small number of simple basic principles that must be assumed without any question. By exhausting every possible combination of these principles, Lull said, we would be able to explore all the knowledge that could be understood by our minds.

Leibniz took Lull's method and tried to apply it to a more formal logic. In "Dissertio de arte combinatoria,"

Leibniz constructed an exhaustive table of all possible combinations of premises and conclusions in traditional syllogism. He eliminated the false syllogisms and identified the number of valid ones. Like Lull, Leibniz was unable to see how restrictive this technique would be.

Leibniz began to try to advance research by making science and language more definite. He employed the model for all reasoning processes in the methods of algebra, where signs take the place of words and remove their ambiguity. He wanted to reduce to a more basic idea all the notions with which the abstract sciences operate. His ultimate goal was to determine combinations of ideas and interpret them by the use of general characters, thereby achieving the strictness and completeness of mathematical reasoning.

Leibniz estimated how long it would take to develop this theory: "I think a few chosen men could turn the trick within five years." Unfortunately, he never did, and toward the end of his life he acknowledged that he had been too distracted by other things.

Although he was to credit Lull with introducing many of the concepts that he was to use, he would become critical years later of Lull's method of logic. In a letter written in 1714 Leibniz commented:

When I was young, I found pleasure in the Lullian art, yet I thought also that I found some defects in it, and I said something about these in a little schoolboyish essay called On the Art of Combinations, published in 1666, and later reprinted without my permission. But I do not readily disdain anything— except the arts of divination, which are nothing but pure cheating—and I have found something valuable, too, in the art of Lully [Lull] and in the *Digestum sapientiae* of the Capuchin, Father Ives, which pleased me greatly because he found a way to apply Lull's generalities to useful particular problems. But it seems to me that Descartes had a profundity of an entirely different level.

Liebniz' brilliant essay on teaching law came to the attention of the Elector's statesman, who requested to have it printed so that a copy could be brought before the Elector. Greatly impressed with his work, the Elector appointed Leibniz the task of revising the code. Within months he was taken into the confidence of the Elector, who used him for a variety of overseas missions. Leibniz had become an official political and business representative of the Elector of Mainz and within a short period became an accomplished diplomat and found himself involved in intrigues and political matters that seriously compromised his scruples. One of his brilliant ideas was to prepare a holy war for the eventual capture and colonization of Egypt. Many years later, Napoleon was quite startled to find that Leibniz had anticipated his campaign.

Leibniz left Germany for the first time at the age of twenty-six and went to Paris. He was sent there on a mission of the court, which was to become of lesser importance than his initiation into the world of science. He had a desire to make intelligible and thinkable the observed connections and relations of things. He had a love for method and order, convinced that such harmony existed in the real world.

Between diplomatic assignments his mathematical education began under the influence of Huygens. Christian Huygens was primarily a physicist but also an accomplished mathematician. He presented Leibniz with a copy of his mathematical essay on the pendulum. Gottfried quickly became impressed with the power of the mathematical approach, and Huygens became his mentor.

It was during this period that Leibniz became fascinated with mechanical contrivances. He was extremely impressed by the idea of a calculating machine and set out to study in detail the works of Pascal and Morland.

Sir Samuel Morland was the son of the Reverend Thomas Morland. Born in 1625 in Berkshire, England, he was educated at Winchester School, the University of Cambridge and later Magdalen College. Although he remained at Cambridge for nearly ten years, he never earned a degree.

Morland started a career in political affairs and soon became the secretary of Oliver Cromwell. Years later he was to become—and remain for a large portion of his life—Master of Mechanics to His Majesty King Charles II. Morland spent most of his time experimenting with mechanical devices.

Sir Samuel Morland replaced "Napier's bones" with discs and became the inventor of an operable multiplier. Around 1666 he invented his arithmetical machine, but did not publish an account of it for several years, when "by the importunity of his very good friends," it was made public. His machine was composed of twelve plates, each of which showed a different part of the mechanism. To operate his machine, a steel pin moved a series of dial plates and small indices. By these means the four aspects of arithmetic were calculated "without charging the memory, disturbing the mind, or exposing the operations to any uncertainty."

The last three years of his life were spent very wretchedly. Poverty and loss of sight compelled him to rely on the charity of friends. Two months before his death the following was recorded:

The Archbishop and myself went to Hammersmith to visit Sir Samuel Morland, who was entirely blind; a very mortifying sight. He showed us his invention of writing, which was very ingenious; also his wooden calendar, which instructed him all by feeling, and other pretty and useful inventions of mills, pumps, etc., and the pump he had erected, that serves water to his garden, and to passengers, with an inscription, and brings from a filthy part of the Thames near it, a most perfect and pure water.

On one of the walls in his house was a stone tablet that read: "Sir Samuel Morland's well, the use of which he freely gives to all persons: hoping that none who shall come after him, will adventure to incur God's displeasure, by denying a cup of cold water (provided at another's cost and not their own) to either neighbour, stranger, passenger, or poor thirsty beggar."

Sir Samuel Morland died on December 30, 1695, in his house at Hammersmith, near the Thames River. He was the king's tinkerer who showed that a mind left alone to choose its path of inquiry can create useful objects to serve man. His arithmetic machine and his other inventions have a place in the history of the calculating machine.

Leibniz studied Morland's and Pascal's various designs and set himself the task of constructing a more perfect and efficient machine. To begin with, he improved Pascal's device by adding a stepped-cylinder to represent the digits 1 through 9.

He almost captured one of the most important aspects of modern-day computing but failed to see its application to mathematics. The binary system so important in present computers was envisioned by Leibniz, but he saw it in terms of religious significance. He saw God represented by 1 and nothing represented by 0. Leibniz used the binary scale as proof that God had created the world (1) out of nothing (0). In fact, he used the binary system as proof to convert the emperor of China to accept a God who could create a universe out of nothing.

But three centuries had to pass before the binary scale was to be found more applicable than the decimal scale in digital computers.

Although Leibniz was amazed by Pascal's calculator, he was not willing to accept it as the ultimate machine. Morland's machine was clumsy and was not always reliable. Leibniz was determined to go beyond these advances and construct a calculator that not only would perform these mathematical feats but would also multiply, divide and extract square roots, and without error: "It is unworthy of excellent men to lose hours like slaves in the labor of calculation which could safely be relegated to anyone else if machines were used."

In 1694, Leibniz built his calculating machine, which was indeed far superior to Pascal's and was the first general-purpose calculating device able to meet the major needs of mathematicians and bookkeepers.

As an attaché for the Elector, Leibniz was sent on a mission to London, where he spent some of his spare time attending meetings of the Royal Society, at which he exhibited his recently completed calculating machine. So impressed was the society with his calculator and other works that he was elected a foreign member of the group before his return to Paris. In the same spirit he was elected to the French Academy of Sciences, becoming the first foreign member of that prestigious institution.

He was never again to attempt the construction of mechanical devices. He returned to his mathematical studies and devoted every spare moment of his time working out some of the elementary formulas that became "the fundamental theorem of the calculus." By the year 1675 he had presented the notation of differential and integral calculus.

Leibniz's remaining forty years were divorced from mathematical and scientific pursuits. Having returned to Germany, he devoted his time to serving the house of Brunswick as historian, librarian and chief adviser.

His popularity spread, and he was invited to accept the

position as custodian of the Vatican Library. But acceptance would have necessitated a conversion to Catholicism, a step Leibniz chose not to undertake. Instead he selected as his major undertaking the reuniting of the Protestant and Catholic churches but found little support for his efforts and was forced to drop this ambitious project.

It was during this time that Leibniz turned to philosophy for his remaining years. He developed the ingenious theory of monads—minute copies of the universe out of which everything in the universe is composed, as a sort of one-in-all, all-in-one. He went on to evolve a basic theorem of optimism: "Everything is for the best in this best of all possible worlds."

Leibniz set out to found the journal *Acta Eruditorum*. His last important contribution came in 1700 in Berlin, where Leibniz organized the Berlin Academy of Sciences and became its first president.

He died in 1716 in Hanover at the age of seventy while still working on the incomplete history of the Brunswick family. A man of numerous interests, his contributions were profound. Some analysts have said that Leibniz would have been a far greater scientist and mathematician if he had not turned to politics and diplomacy. For these interpreters of history Gottfried Leibniz had wasted too much of his time with the unimportant issues of the day, time that could have been more profitably turned toward science.

But for Leibniz specialization was short-lived. In his long life he showed genius in just about every endeavor he attempted. His mind was encyclopaedic, and his learning was wedded to practical and even humanitarian activities. He united qualities rarely found together: he was both practical and speculative, a master of induction and generalization as well as an erudite and patient scholar, a believer and daring innovator.

The master architect of German philosophy in the seventeenth century, Leibniz will be remembered for his many achievements. In the light of his other accomplishments his all-purpose calculating machine was one of his lesser feats. But it should be remembered that a man of lesser brilliance would certainly have found a place in history had he merely built a calculating machine on the scale that Leibniz did.

As we trace the evolution of the modern-day computer, we see that Leibniz' efforts were necessary and were a further advance in the mechanization of the calculator.

3

Charles Babbage

The Unsung Hero of the Computer

Charles Babbage, when asked to write his own life story, replied that he had no desire to do it while he had the strength and means to do more important work. Some people, he said, write their lives to save themselves from boredom, while others fear that their true story will never be told. He belonged to neither of these groups. When asked to prepare an autobiography, he merely gathered together a list of his publications and sent them on.

Babbage is a representative of that unusual group of talented engineers and scientists who have altered the course of human events. Unfortunately, most of his life was spent in an unsuccessful attempt to construct a machine that his friends and associates thought to be completely ridiculous. They urged him to abandon it and save his money, his time and his reputation.

Like so many productive pioneers, Babbage was born at the wrong time, perhaps a hundred years too soon. From the very start, controversy dominated his life. The place and

date of his birth is no exception. As an old man he sent a
letter to the Statistical Society: "You may inform the French
gentleman who made this enquiry that the place of my birth
was London and the year was 1792." But he had erred.
Either he never knew the truth about his birth, or the truth
was slipping away as a result of advancing years. He was
born on December 26, 1791, in Totnes, Devonshire, England.

It was an exciting time in history. The King of France,
Louis XVI, and his Queen, Marie Antoinette, accompanied
by their children, were trapped trying to escape from
France. Their identity revealed by a postmaster, they were
arrested, humiliated and returned to Paris.

There were many who sympathized with the French Revolution. Dr. Joseph Priestly, discoverer of oxygen and friend
of the people of France, had planned a public banquet to
celebrate the anniversary of the fall of the Bastille, July 14.
But his dinner party never was held, for his political ene-
mies broke into his house, destroyed his personal and household belongings, his instruments and, most important, his
collection of research findings.

This was the world that Babbage entered. He was denied
the right to a relative's large estate, the result of a sudden
act of the British Parliament in which rights of inheritance
were reorganized.

"I think myself very fortunate in thus having escaped the
temptations as well as the duties of wealth," he wrote. "Even
if I had resisted the former, long years of continuous intellectual labour would have entirely prevented me from fulfilling the latter." This was to change, however, when
Babbage reached the age of thirty-six. When his father died
he inherited the enormous sum of about £100,000.

From the very beginning, Charles showed a great desire
to inquire into the causes of things that astonish children's

minds. Seventy years later he was to say in jest that he could trace his ancestry to the prehistoric flint workers because of his "inveterate habit of contriving tools." Upon receiving a new toy Charles would ask: "Mamma, what is inside of it?" Then he would carefully proceed to dissect it and figure out how it was constructed and what made it work.

He was once caught attempting to prove the existence of the devil by drawing on the floor of his house a circle with his own blood and irreverently spouting the Lord's Prayer backward.

Babbage overcame toothaches by reading *Don Quixote*. He attempted to play with the supernatural spirits by arranging with another boy that whoever died first would appear to the survivor and some years later stayed awake all night to await the apparition of his friend, recently deceased.

At Trinity College, Cambridge, he proved to be an undisciplined student constantly puzzling his tutors. He was often annoyed to find that he knew more than his masters. But in spite of his outward displays of rebellion he was already on his way to absorbing the advanced theories of mathematics.

Charles became an expert in chess, beating every challenger that he confronted. He formed a ghost club to collect any reliable evidence to support the existence of the supernatural. Sailing became a hobby of his, not from a "love of sea" point of view, but from the intellectual aspects involved in the manipulation of a floating vessel.

With several others Babbage formed an Analytical Society to present and discuss original papers on mathematics and to interest people in translating the works of several foreign mathematicians into English. Despite considerable opposition, the Society fought hard to put "English mathematicians on an equal basis with their continental rivals."

During Babbage's stay at Cambridge his studies led him

into a critical examination of the logarithmic tables used in making accurate calculations. The time and labor to construct these tables were great.

During one of his freer moments, Babbage was contemplating a problem while sitting in a room of the Analytical Society. One of his friends entered and seeing Charles appearing to be in a far-off world, asked, "Well, Babbage, what are you dreaming about?" Pointing to some tables of logarithms, Charles replied, "I am thinking that all these tables might be calculated by machinery." He was forever calling to the attention of scientific societies the number and importance of errors in astronomical tables and calculations.

Following graduation he returned home to begin his work in sketching a machine by which all mathematical tables could be computed by one uniform process. He was convinced that it was technically feasible to construct a machine to compute by successful differences and set type for mathematical tables. His first model consisted of ninety-six wheels and twenty-four axes, later to be reduced to eighteen wheels and three axes.

His interest in mathematical perfection caused him to challenge the accuracy of words in Alfred Tennyson's couplet "The Vision of Sin." Babbage wrote to the poet:

"Every minute dies a man,/Every minute one is born": I need hardly point out to you that this calculation would tend to keep the sum total of the world's population in a state of perpetual equipoise, whereas it is a well-known fact that the said sum total is constantly on the increase. I would therefore take the liberty of suggesting that in the next edition of your excellent poem the erroneous calculation to which I refer should be corrected as follows: "Every moment dies a man/ And one and a sixteenth is born"! I may add that the exact figure is 1.167, but something must, of course, be conceded to the laws of metre.

Interestingly enough, all editions of the poem up to 1850 read: "Every minute dies a man/Every minute one is born." All editions after 1850 read: "Every moment dies a man/ Every moment one is born."

When Charles was twenty-three years old he married a twenty-two-year-old named Georgiana. His marital life was full of grief and loss. Georgiana allowed her husband to dominate her life completely. When his children arrived, Charles would often bury himself in his work, rarely allowing the cries of his offspring to interfere with his concentration. Their marriage produced eight children in thirteen years and terminated with the untimely death of his wife at the age of thirty-five. Only three sons of his eight children survived to adulthood.

Babbage's first academic post was in 1816 as professor of mathematics at the East India College, for a salary of £500 a year. This was followed by the chair of mathematics at the University of Edinburgh.

Babbage was in a hurry to start applying his ideas for mechanizing the logarithm tables. In 1822, at the age of thirty, he announced at a meeting of the Royal Astronomical Society that he had embarked on the construction of a machine for the purpose of calculating tables. His power of influence was great and so was his conviction of success. "I have taken the method of differences as the principle on which my machinery is founded, and in the engine which is just finished I have limited myself to two orders of differences. With this machine I have repeatedly constructed tables of square and triangular numbers as well as a table from the singular formula $x^2 + x + 41$ which comprises amongst its terms so many prime numbers."

His paper "Observations on the Application of Machinery to the Computation of Mathematical Tables" was received with such acclaim that he was presented the first gold medal award by the Astronomical Society.

Now Babbage was determined to arouse the interest of the famous and prestigious Royal Society. On July 3, 1822, he continued his plea in a letter to Sir Humphrey Davy, President of the Royal Society, stating that the "intolerable labour and fatiguing monotony of a continued repetition of similar mathematical calculations had first excited his desire and afterwards suggested the idea of a machine which by the aid of gravity or any other moving power should become a substitute for one of the lowest occupations of the human intellect."

On April 1, 1823, the treasury of the Royal Society received a letter from its president, requesting that the council take into consideration Babbage's plan, which had been submitted to the government, for applying machinery to the purposes of calculating and printing mathematical tables.

Babbage pleaded that, though he had arrived at a point where success was no longer doubtful, "it could be attained only at a very considerable expense, which would not probably be replaced by the works it might produce for a long period of time; and which is an undertaking I should feel unwilling to commence, as altogether foreign to my habits and pursuits."

The council of the Royal Society appointed a twelve-man committee to consider this proposal. On May 1, 1823, with only one dissenting opinion, the council reported that "it appears Mr. Babbage has displayed great talents and ingenuity in the construction of his machine for computation, which the committee think fully adequate to the attainment of the objects proposed by the inventor, and that they consider Mr. Babbage as highly deserving of public encouragement in the prosecution of his arduous undertaking."

This report and accompanying documents of request

were printed and laid before Parliament. Two months passed, and the treasury granted Babbage £1,500 "to enable him to bring his invention to perfection in the manner recommended."

Initial excitement began to turn into concern when a misunderstanding developed between Babbage and the British Government. This confusion lasted for twenty years and is rather odd in light of present-day thinking. Babbage regarded the machine that he was going to build as the property of the Government of England. They in turn understood it to be his. He had received the first advance of funds with the expectation that all necessary monies would be furnished to complete the project. They seemed to have regarded the advance as a temporary assistance, given to a man of genius for the purpose of enabling him to complete an invention that would be of great public value.

At about this time, one of the most honored academic positions was offered to Babbage, the Lucasion Professorship at Cambridge, the chair once held by Isaac Newton. At the outset he hesitated, thinking it would interfere with his work on the Difference Engine, but he was persuaded to change his mind and held the position for a few years. Many years later, Charles was to say that this was the only honor he received in his own country.

After patiently waiting the support of the Government, funds were now available, and Charles set about planning his Difference Engine.

It is still not clear where he got his basic ideas for his engine; he was not alone in trying to work out ways to design such a machine. In 1782, thirty years before Babbage built his Difference Engine, Johann Helfrich von Müller invented and built a calculating machine. This device required that problems had to be set up by hand and that after revolving the crank that appeared in the top center, the solution was read from the number wheels. His machine was a true calculator that could multiply by repeated addition and divide by repeated subtraction. The wheels and gears found in Müller's machine appear to be quite similar to those used by Babbage years later.

In 1786, Müller conceived the idea for mechanical differencing. According to Cajori in his *History of Mathematics,* "the first idea of automatic engine calculating by the aid of functional differences of various orders goes back to J. H. Müller, 1786." Müller should rightfully share with Charles Babbage some of the credit for many of the concepts found in the difference engine.

It is not clear whether Babbage knew of Müller's work or if he discovered his principles during independent research and observation. The suspicion remains that Babbage did know of Müller's device but never, at least in writing, credited him for his contribution.

Babbage studied the inventions of several predecessors, but few were as impressive as the work of Charles Mahon, Third Earl Stanhope. Mahon's arithmetic machine was of such mechanical ingenuity that Babbage used it as a foundation for the development of his own calculating engine.

Mahon came from a celebrated English family. His grandfather was a commander in the War of the Spanish Succession and Prime Minister of King George I, and his grandmother was the daughter of Governor Pitt. His father, devoting most of his life to science, became a prominent contributor of knowledge rather than a member of politics and society. It was his wife, Grisel, who maintained the charm necessary for retaining the family appearances in social circles.

Born in London on August 3, 1753, Charles Mahon was their only surviving child. Although he attended Eton, his family was determined that the English climate would not take a toll on his health, and he was sent to school in Geneva. His tutor at Eton wrote to his parents: "He has an excellent understanding, though he does not seem to relish books. He has an excellent heart and disposition. Never have I known his equal for so early a sense of honour."

In Switzerland the education was extremely good, and the nation could boast of a public library of forty thousand volumes while not a single free library existed in Great Britain.

Under the influence of Georges Le Sage, Mahon applied himself to geometry, mechanics and philosophy, and by the age of seventeen he won the prize offered by the Swedish Academy for the best essay on the construction of the pendulum. One of the major results of his ten years of education in Switzerland was his scorn of luxury and display so dominant among the students of England. Geneva instilled in him a devotion to science and philosophy and, above all, a love of civil liberty.

When he was nineteen he was elected a fellow of the Royal Society, and at the age of twenty-one young Mahon, with his family, went to live in Paris. He immediately became friends of some of the more learned men, who were well aware of the skill that had brought him his prize from the Swedish Academy.

In July, 1774, the family returned to England. It was an important year in Charles's life. In September he proposed to Lady Hester Pitt, his second cousin, and they were married on December 19, 1777.

After an unsuccesful attempt at politics, Charles decided to devote most of his time to scientific pursuits.

In 1777, the year after the conclusion of the American War of Independence, Charles Mahon invented two calculating "arithmetic machines." The first, "by means of dial-plates and small indices, moveable with a steel pin, performed with undeviating accuracy" complicated sums of addition and subtraction. The second solved problems in multiplication and division, "without possibility of mistake," by the simple revolution of a small winch. It was half the size of a common table writing-desk, and "what appears very singular and surprising to every spectator of this machine is that in working division, if the operator be inattentive to his business and thereby attempts to turn the handle a single revolution more than he ought, he is instantly admonished of his error by the springing up of a small ivory ball."

Of great importance was the use by Mahon's machine of gear wheels and a tens-carrying device. The machine contained a series of toothed wheels, having wide faces bearing ten very long teeth; the first one, reaching clear across the face, representing 9, the next tooth being one-ninth shorter, the next one-eighth shorter, and so on. To add 9, the toothed wheel was moved along so that the nine teeth would engage; to add 8, it was moved along so that eight would engage.

Mahon's life was full of invention. His "Stanhope Demonstrator" was the first crude attempt of a kind of inductive-logic machine. He took out a patent for "constructing ships and vessels and moving them with help of sails, and against wind, waves, current or tide"—years before Fulton had begun to work on the steamboat.

Another important invention was in the art of printing, especially in the process of stereotyping. He said, "The art of printing has contributed so eminently to the civilization of society that men of science ought to do their best to improve it. To every man who is fully sensible of the importance of diffusing knowledge the dearness of books must be a subject of considerable regret. This evil arises in a great measure from the expense and risk to which publishers are liable." If the publisher printed too many copies it was a loss; if too few, he must incur the expense of a new edition.

Among Mahon's minor innovations the most useful was the small but powerful lens that bears his name. He also developed a monochord for tuning musical instruments, a system for fireproofing buildings, certain improvements in canal locks, better uses of mortar in construction, and roof covering materials made of tar, chalk and fine sand. Together with Benjamin Franklin he worked on several theories of electricity.

Mahon died on December 17, 1816. "I desire," he had written in his will, "that my funeral may be conducted without the least ostentation, as if I were to die a very poor man." His wishes were respected. There was neither hearse nor mourning coach, but the body was carried to the grave.

He was one of the most striking personalities of his time. As one of his closest friends said: "On the whole, if a scrupulous regard to truth prevents me from pronouncing my friend and benefactor, without some qualification, a wise, a good, or an amiable character, none can deny that he was an ingenious, diligent, disinterested, useful, and remarkable man."

In the Stanhope Chapel, where he is buried, an inscription reads:

He was endowed with great powers of mind, which he devoted zealously and disinterestedly to political and general science, to the diffusion of knowledge and to the extension of civil and religious liberty. He requested, by his will, to be buried "as a very poor man," and in the spirit of that injunction this plain tablet is inscribed to his memory. His understanding and integrity would have raised him to the notice of his fellow-countrymen even if his lot had been cast in the humblest condition of life.

Lord Stanhope's passion in life was the service of his countrymen. It was his misfortune that while few men have worked so ardently and so consistently for mankind, few have reaped so little gratitude.

Although Babbage's work would have been carried out, stumbling on Mahon's inventions made it easier for him to master some of the more difficult mechanisms that he was later to employ.

When Babbage began to work on his Difference Engine, he accepted no money for his own efforts, because of his belief that taking such funds would be a direct violation of his contract with the Government. The Government followed his progress and continued to furnish him with more funds. At Government expense large sums of money were set aside for his research, and a fireproof building with a workshop was built on leased land adjoining his home.

However in 1833, while arrangements were being made to move the engine and his other materials to the workshop, Babbage was faced with a dilemma. There had been delays in salary payments to his engineer, Joseph Clement. One day Clement refused to continue his work and dismissed all his employees. Although he eventually allowed the engine to be transferred to the new quarters, he refused to let Babbage have access to his tools, which were specially constructed over a period of eight years at Babbage's and the Government's expense. After breaking with Clement, Babbage was never again to carry out his projects beyond the stage of drawings.

It seems clear now that Babbage made two major miscalculations. First, constructing the Difference Engine would have cost about fifty times more money than he was given. Second, he needed about two tons of novel brass, steel and pewter clockwork, which had to be made since it was not available in his day.

During this period, the Countess of Wilton and the Duke

of Wellington paid a call at Babbage's home to see the Difference Engine. The Countess asked what he considered his biggest obstacle in developing the engine. Charles replied that his greatest barrier was not that of "contriving mechanisms to execute each individual movement . . . but it really arose from the almost innumerable combinations amongst all these contrivances." He then proceeded to draw an analogy of his problems to those of a military leader commanding a large army in battle. The Duke of Wellington, the noted warrior, was in full accord with Charles' comparison.

Babbage's Difference Engine was created to calculate and print mathematical tables. The eleventh edition of the Encyclopaedia Britannica, long out of print, describes his Engine:

Imagine a number of striking clocks placed in a row, each with only an hour hand, and with only the striking apparatus retained. Let the hand of the first clock be turned. As it comes opposite a number on the dial the clock strikes that number of times. Let this clock be connected with the second in such a manner that by each stroke of the first the hand of the second is moved from one number to the next, but can only strike when the first comes to rest. If the second hand stands at 5 and the first strikes 3, then when this is done the second will strike 8; the second will act similarly on the third, and so on. Let there be four such clocks with hands set to the numbers 6, 6, 1, 0 respectively. Now set the third clock striking 1, this sets the hand of the fourth clock to 1; strike the second (6), this puts the third to 7 and the fourth to 8. Next strike the first (6); this moves the other hands to 12, 19, 27 respectively, and now repeat the striking of the first. The hand of the fourth clock will then give in succession the numbers 1, 8, 27, 63, etc., being the cubes of the natural numbers. The numbers thus obtained on the last dial will have the differences given by those shown in succession on the dial before it, their differences by the next, and so on till we come to the constant difference on the first dial. A function

$$y = a + bx + cx^2 + dx^3 + ex^4$$

gives, on increasing x always by unity, a set of values for which the fourth difference is constant. We can, by an arrangement like the above, with five clocks calculate y for x = 1, 2, 3 . . . to any extent.

This is the principle of Babbage's difference machine. The clock dials have to be replaced by a series of dials, . . . and an arrangement has to be made to drive the whole by turning one handle by hand or some other power. Imagine further that with the last clock is connected a kind of typewriter which prints the number, or, better, impresses the number in a soft substance from which a stereotype casting can be taken, and we have a machine which, when once set for a given formula like the above, will automatically print, or prepare stereotype plates for the printing of, tables of the function without any copying or typesetting, thus excluding all possibility of errors.

The work on the Difference Engine ceased, and a year went by before Babbage returned to his engine. But by this time he had developed an entirely new principle for a machine that was to be an improvement over the Difference Engine. It would be an engine—the Analytical Engine, he called it—that would be easier to construct, have greater versatility and operate at a faster rate. While the Difference Engine was designed to perform a limited set of operations, the Analytical Engine was designed to perform all arithmetic calculations and to combine such operations to solve any conceivable arithmetic problem.

In fact, except for the relatively slow speed of sixty additions per minute, Babbage's Analytical Engine had the sort of flexibility that present-day computers have.

His machine was composed of four parts. He had a "store," in which the numerical data involved in the calculation would be kept. It would be composed of columns of wheels, and on each would be ten engraved digits. In all he made thousands of mechanical drawings to describe his "stores."

The "mill" was the second basic part of his computer.

Arithmetical calculations were made through the rotation of its gears and wheels.

The third device was a series of gears and levers that transferred numbers back and forth between the "store" and the "mill."

His last device was for moving numerical information in and out.

Incorporating the method of punch cards used by the textile inventor Jacquard, Babbage was able to set up a pattern of holes that corresponded to mathematical symbols. "We may say most aptly that the Analytical Engine weaves algebraic patterns just as the Jacquard-loom weaves flowers and leaves."

Prior machines required manual intervention at various steps. The Analytical Engine had eliminated the need for most of the physical operations. "This engine surpasses its predecessors, both in the extent of the calculations which it can perform, in the facility, certainty and accuracy with which it can effect them, and in the absence of all necessity for the intervention of human intelligence during the performance of its calculations."

Unfortunately, government monies had ceased. Babbage, on the other hand, had used his own capital, for which he was not reimbursed. He soon found that his efforts were no longer being encouraged by members of the society, government and the scientific men of Europe who passed themselves off as philosophical mechanists. After all his work he received a letter informing him that "ultimate success appeared so problematical and the expense so large and so utterly incapable of being calculated, that the government would not be justified in taking upon itself any further liability."

Had this note been received shortly after commencing on this project, Babbage would have interpreted it as a lie. But in twenty years it can become obvious that a dream in one's mind that seems to work doesn't necessarily become a workable machine when one attempts to build it. Even Babbage was to learn that there was truth in this. With all his zeal and excitement over this undertaking he had grossly underestimated the practical and engineering difficulties. While the Royal Society and the British Government were debating whether to cut off his funds, he was to say that "while an inventor might not be far from the truth in having imagined an instrument possessing the most remarkable powers of calculation, he would be very remote from the attainment of its execution and probably spend months or years or even life itself in vainly attempting to surmount the difficulties of some few of its successive stages."

In 1841, Babbage himself was thoroughly frustrated with his lack of progress. After formulating the general procedure for constructing his machine, the first step was to divide the plan into basic units and develop the necessary mechanisms to permit a smooth flow of operation. From one part to another there was a need for continuous subdivisions, allowing one gear to fit another, then another, and so on. It became obvious that special mechanisms had to be invented from basic drawing plans. For example, in addition it would be necessary to have a combination of mechanisms for adding digits, and another set of gears for carrying over the tens.

Not only did the learned scientists and professionals attack Babbage, but he was indirectly challenged by some of the more popular writers of the day. Edgar Allan Poe, in "The Chess Player of Maelzel," wrote:

What should we think of a mechanism of wood and metal which not only computes astronomical and nautical tables to any degree of accuracy, but can also guarantee the mathematical accuracy of its observations by its own power of correcting possible errors? What are we to think of a mechanism that can not only do this, but can actually print the results of its complicated calculations as soon as they are obtained and perform all this without any intervention of human intelligence.

During this period of rejection Babbage was faced with numerous personal tragedies. In 1827 he wrote:

I lost my father, my wife and two children. My family acting on the advice of my medical friends, urged me to travel abroad for six or twelve months. It was thought necessary that I should be accompanied by a servant in case of illness or accident. I objected to this on the ground that I was but just able to take care of myself, and that a servant would (to me) be a great encumbrance. To satisfy, however, a mother's anxiety I proposed to take with me one of my own workmen if he like to accompany me as an attendant.

While abroad Babbage met numerous mathematicians, mechanics and members of aristocracy. For these encounters he relied on what he could pull out of his bag of scientific gadgets. "It is always advantageous for a traveller to carry with him anything of use in science or in art if it is of a portable nature, and still more so if it has also the advantages of novelty."

One of his most cherished objects was a collection of gold buttons stamped by steel discs with ruled parallel lines 4/10,000 of an inch apart, designed by the comptroller of the mint. The rainbow panorama these buttons produced provided a perfect introduction for a conversation.

Babbage once designed a tick-tack-toe machine to help raise money for his serious efforts. The machine had figures of a lamb, a cock and two children each in turn clapping, crying and crowing. The mechanism was a slow but real automatic system:

It occurred to me that if an automaton were made to play this game, it might be surrounded with such attractive circumstances that a very popular and profitable exhibition might be produced.

When it became known that an automaton could beat not merely children but even papa and mamma at a child's game, it seemed not unreasonable to expect that every child who heard of it would ask mamma to see it. On the other hand, every mamma, and some few papas, who heard of it would doubtless take their children to so singular an interesting a sight. I resolved, on my return to London, to make inquiries as to the relative productiveness of the various exhibitions of recent years, and also to obtain some rough estimate of the probable time it would take to construct the automaton, as well as some approximation to the expense.

Charles was advised to abandon this venture since it probably would have been a financial failure in trying to compete with General Tom Thumb, the favorite game of the day.

Partially rehabilitated, he returned to England at the end of 1828 but chose to live apart from his children. His three sons were to grow up with Babbage's mother.

Charles's life was soon to become enriched through a relationship with Lady Lovelace, the only legitimate child of Lord Byron. At the age of fifteen she began to display her interests in mathematics by teaching herself geometry. Perhaps, because of this love for figures, or perhaps because of her need for a father-image (Lord Byron deserted his wife only a few months after her birth), she became magnetically attracted to Charles Babbage, and he to her.

Ironically, Lord Byron and Babbage were as different in personality and interests as any two people could be. Charles was attempting to mechanize a system to make life freer from the mundane effort of simple calculations while the lord-poet wished to halt the progress toward mechanization of work. Byron sympathized with the Luddites, who were trying to destroy the mechanization of the textile industry out of fear of unemployment. In 1816, Byron wrote to Thomas Moore: "Are you near the Luddites? By the Lord! if there's a row, I'll be among ye! How go on the weavers—the breakers of frames—the Lutherans of politics —the reformers?"

In a conflicting world of poetry and mathematics, Lady Lovelace was able to retain her flowering love for the word, for peace and for people. The editor of *The Examiner* once described her:

She was thoroughly original, and her genius, for genius she possessed, was not poetic, but metaphysical and mathematical, her mind having been in constant practice of investigation, and this with the vigorous exactness. With an understanding thoroughly masculine in solidity, grasp and firmness, Lady Lovelace had all the delicacies of the most refined female character. Her manners, her tastes, her accomplishments, were feminine in the nicest sense of the word; and the superficial observer would never have divined the strength and the knowledge that lay hidden under the womanly graces. Proportionate to her distaste for the frivolous and the commonplace was her enjoyment of true intellectual society, and eagerly she sought the acquaintance of all who were distinguished in science, art and literature.

But it was her romance with mathematics that brought Lady Lovelace and Babbage closely together. In 1842, at the age of twenty-seven, she translated from the French a paper dealing with the Analytical Engine. Babbage was delighted but questioned why she bothered to translate the paper instead of creating something original. The idea was appealing to Lady Lovelace, and she began to add some notes to the original manuscript. By the time she was finished with her additions to the French work, her notes had accumulated to about three times the length of the original paper.

Charles was to pay her tribute for these efforts: "These

memoires, taken together, furnish, to those who are capable of understanding the reasoning, a complete demonstration that the whole of the developments and operations of analysis are now capable of being executed by machinery."

By 1843, Lady Lovelace had mastered Babbage's plans for his engine. She studied his ideas, made extensive notes and used her influence to get others interested in them. One of her memos revealed her depth of involvement: "I am working very hard for you; like the Devil in fact; (which perhaps I am)." Her energetic attempts to make Babbage's work more understandable to the public led her to evolve a form of arithmetic called binary, which used ones and zeros. Eventually, other people would develop binary as the basic mathematics of the computer.

Her affection and remarkable attachment to Babbage came to an abrupt and pathetic end—strangely enough at the same age as her father, Lord Byron. The illness that confronted Lady Lovelace drove her to acts of violence. Her bedroom had to be lined with mattresses to prevent injury for she was frequently found throwing herself about in severe agony. The pain was to stop in 1852 when she was thirty-six. She was buried beside her father in the Byron vault.

Thus ended one of the most remarkable and rewarding relationships between man and woman. In a period of so few years Lady Lovelace had experienced the widest range of love and activity. A brilliant interpreter of Babbage, she completely understood and appreciated his machine and is credited with describing his work with such clarity to the community of intellectuals that even Babbage was to admit that her presentation of his work was clearer than his own.

Lady Lovelace was one of the few people in her time to be able freely to discuss and interpret Babbage's two engines. She anticipated many of the terms and ideas used today. Her insights were remarkable: "Supposing, for instance, that the fundamental relations of pitched sounds in the science of harmony and of musical composition were susceptible of such expression and adaptations, the engine might compose elaborate and scientific pieces of music of any degree of complexity or extent."

She felt that the Analytical Engine was to the Difference Engine as analysis was to the simpler arithmetic. The Analytical Engine could add, subtract, multiply or divide, while the Difference Engine was merely an adding machine that was able to tabulate but could not develop relationships between numbers.

This champion of Babbage had made her mark, not only on Charles but on the rest of the world. Her death marked the end of a brilliant life and concluded another chapter in Babbage's continual battle with disaster and sorrow.

Life went on, and Charles Babbage remained in the center of controversy. One of his most critical writings was the "Ninth Bridgewater Treatise." The Earl of Bridgewater had set aside a fund for the writing of a series of works "On the Power, Wisdom, and Goodness of God, as Manifested in the Creation." Babbage, in adding a ninth treatise at his own expense, developed the same subject but in a different way. He argued against the prejudice, which he felt was introduced in the first of the volumes, that the pursuits of science, and of mathematics, were unfavorable to religion.

Babbage believed that many of the most respected clergy had presented God as constantly interfering to change the laws that He had previously ordained, thus, said Babbage, denying by implication to Him the highest attribute of omnipresence.

It was the machine that Charles used to illustrate his point. He stated that the Difference Engine could be used to demonstrate those laws, which could operate for ages and only after a long period of time would be altered into different laws.

His argument was pursuasive. He concluded that the Creator's works were proof of the existence of a Supreme Being. And the greatest of these efforts was the creation of man.

For man, the greatest of virtues was truth:

And the true value of the Christian religion rested, not upon speculative views of the Creator, which must necessarily be different in each individual, according to the extent of the knowledge of the finite being, who employed his own feeble powers in contemplating the infinite; it rested on those doctrines of kindness and benevolence which that religion claimed and enforced, not merely in favour of man himself, but of every creature susceptible of pain or happiness.

The only way to preserve a future existence, according to Babbage, was by developing and possessing memory. "If memory be absolutely destroyed, our personal identity is lost."

Although Babbage never used the word, he was a strong believer in the reincarnation of life. He said on several occasions that in some future condition we might awaken to find ourselves believing that we had existed not in one former condition but in many. The type of return would be a direct reward or punishment for the way we had conducted ourselves in earlier periods.

His most startling pronouncement about creativity dealt with the dominance and superiority of man over other animals. Man, he stated, through his senses, possessed five sources of knowledge. Man thinks of himself as the greatest achievement of God; but, asked Babbage, might it not be possible that man is the lowest achievement of God? On the basis of what was known of other species in the world, Babbage questioned the possible superiority of different animals. If other animals possessed senses of a different nature from ours, it could scarcely be possible that man would ever be aware of it. "Yet those animals, having other sources of information and of pleasure, might, though despised by us, yet enjoy a corporeal as well as an intellectual existence far higher than our own."

According to his critique, "he thought of God as a Programmer." Babbage's arguments covered many areas of religion. He had tried in his treatise to show the possible origin of miracles, in a world otherwise controlled under God by orderly natural law.

But Babbage was not to become a religious philosopher. His passions in life were the Difference and Analytical Engines.

One of his closest friends was Baron Plana, a mathematician of considerable reputation. Excited by and still supportive of Babbage's work and ideas, Plana invited Charles to Turin, to present his studies to some of Italy's greatest mathematicians and engineers. Italy took Babbage in as if he were native-born. A private audience was arranged with King Charles Albert.

Following his stay in Italy, Babbage traveled through France and visited Lyons in order to see how silk and other textiles were manufactured. In particular he wanted to see the loom on which "that admirable specimen of fine art, the portrait of Jacquard, was woven." He purchased a copy of this outstanding portrait and brought it back to Italy to present to the Queen.

Once again an audience was arranged, and Babbage entered the royal suite carrying a large case inside of which was the portrait of Jacquard. Charles opened the package in order to show it to the King, who was pleased to allow Her Majesty to accept such a handsome gift. When Babbage began to rewrap the portrait, some of the silver wrapping started to fly from the box. He tried to reach for them, the King tried, but within seconds the sheets were on the floor. Then there followed a scene that Babbage was never to forget—the King and he were crawling on their knees, picking up silver paper: "I perceived that the heel of royalty had come into contact with the toe of philosophy. A comic yet kindly smile beamed upon the countenance of the King, whilst an irrepressible but not irreverent one, lightened up my own."

This encounter endeared Babbage to King Albert, and upon Babbage's return to London in 1841, a new popularity allowed him to attempt to renegotiate with the British Government and the Prime Minister, Robert Peel. For this important meeting, in which he hoped to persuade the Government to release more monies for his work, Babbage prepared a detailed statement of his past activities:

Of course, when I undertook to give the invention of the calculating engine to the Government, and to superintend its construction, there must have been an implied understanding that I should carry it on to its termination. I entered upon that undertaking believing that two, or at the utmost three, years would complete it. The better part of my life has now been spent on that machine, and no progress whatever having been made since 1834, that undertaking may possibly be considered by the Government as still subsisting. I am therefore naturally very anxious that this state of uncertainty should be put an end to as soon as possible.

In a personal interview with Sir Robert, Babbage argued that the failure to complete the whole engine as originally planned was not his fault. Putting the blame where he believed it to be, he stated that the Government was solely and entirely at fault, and he requested a fair dealing in the matter. He wanted some compensation for his twelve years of devoted work. But primarily he was seeking further support to build the Difference Engine.

During the meeting with Peel some words were exchanged that angered the Prime Minister, and later Babbage recorded: "Finding Sir R. Peel unwilling to admit that I had any claim, I merely remarked that I considered myself as having been treated with great injustice, but that as he seemed to be of a different opinion, I could not help myself, on which I got up and wished him good morning."

Charles wanted this interview for two reasons. He wished to convince the Prime Minister that a Difference Engine could be built and that it was also within reason to expect that an Analytical Engine might be completed within a few years. But the reaction of Sir Robert appeared to destroy these possibilities.

As a token of respect to Babbage and the years he had spent in this apparently unproductive venture, the Prime Minister offered him a baronetcy, which Babbage declined.

A depressed genius, Babbage devoted the six years following the termination of governmental support to the construction of his Analytical Engine. It was during this period that he finally mastered the theories behind the Engine and uncovered and identified the unlimited capabilities that such a machine could provide. Without any financial support from the Government, Babbage continued to work at his own expense, depleting his savings little by little. He hired draughtsmen and other workers to be employed in his house and continued the experimentation that would lead to his greatest evidence of genius.

It was during this period of his life that Charles became a prolific writer and speech maker. One of his more important papers dealt with the subject of taxation. The reaction of many was profound. From the noted English writer Charles Dickens he received a congratulatory letter:

Devonshire Terrace, 26 Feb. 1848

My dear Sir,

Pray let me thank you for your pamphlet. I confess that I am one of the inconvenient grumblers, and that I doubt the present or future existence of any Government in England strong enough to convert the people to your income tax principles, but I do not the less appreciate the ability with which you advocate them, nor am I the less gratified by any mark of your remembrance.

Faithfully yours always,
Charles Dickens

There was little design work left for Babbage to complete on the Analytical Engine by now, and so he turned his attention to designing a series of plans for his Difference Engine No. 2. In reality, this was to be an improvement over the Difference Engine No. 1 because of the inclusion of simplified devices that were to be used in his Analytical Engine.

By 1852 his friend the Earl of Rosse was elected to the presidency of the Royal Society. As a follower of Charles's work, Rosse asked Babbage whether he was willing to turn over his recent drawings if the Government would have the machine constructed.

Babbage replied, "My feeling was, after the sad sacrifice of the past, that I ought not to think of sacrificing any further portion of my life upon the subject. . . . If, however, they chose to have the Difference Engine made, I was ready to give them the whole of the drawings and notations."

Again, bad luck struck Babbage. While preparing a letter to the Prime Minister discussing the likelihood of submitting his work, there was a change of Government and once again Charles's dream was abruptly brought to a halt. A defeated man, he said, "Science was weighed against gold by a new standard, and it was resolved to proceed no further." He was thoroughly disgusted with the way he had been treated and denied his true place in society. "I cannot but feel that whilst the public has already derived advantage from my labours, I have myself only experienced loss and neglect," he wrote in a letter to Lord Derby, the new Prime Minister.

The matter was then referred to Benjamin Disraeli, who was the Chancellor of the Exchequer. Disraeli soon learned that over the years Babbage had never completed building a machine. Therefore his reaction was short and to the point. Disraeli stated that a Difference Engine as described by Babbage would cost the Government too much money, that the costs would be impossible to determine, and that it would in all probability never lead to a worthwhile tool.

The Chancellor's statement, cold and quick, released from Babbage a series of equally cold and quick retorts. Of his Difference Engine: it "can not only calculate the millions the ex-Chancellor squandered, but it can deal with the smallest quantities, nay it feels even for zeros." In another notation: "It may be necessary to explain to the unmathematical reader and to the ex-Chancellor of the Exchequer that *impossible quantities* in algebra are something like mares' nests in ordinary life."

Even before his falling out with the British over their lack of support for building his Difference Engine, Babbage was a leading protagonist in criticizing the Government for neglecting its science and researchers. Outraged with the knowledge that few Englishmen pursued science, he led an all-out attack on the deteriorating science educational system and demanded Government support for its scientists.

His book *Reflections on the Decline of Science in England and on Some of Its Causes* evoked much criticism and resentment. The book claimed that England, although a power in mechanical and manufacturing areas, was far below other countries in the more important areas of pure science. The educational system was found to be ineffective, and

Babbage outlined a plan for revising the universities. Many years were to pass before some of his reforms were introduced.

Babbage was also a strong advocate of introducing scientists into the leadership of Government: "It is of the very nature of knowledge that the recondite and apparently useless acquisition of today becomes part of the popular food of a succeeding generation."

Within his own professional society circles he was a maverick. He attacked the intrigues of the Royal Society, considered the officers unqualified and even claimed that the president was selected more on the basis of rank than on his professional credentials. Babbage's sweeping reforms included establishing a set of requirements for admission to the learned society to be based on publication of scientific papers, setting up democratic procedures for electing members and allowing free and open debate on the operations of the Society.

At this time, his son Henry was returning to England after serving in the army for ten years. In earlier periods, his father seemed to have little to do with his children for they were not his intellectual equivalent.

But now the senior Babbage was sixty-three and had had many years to contemplate his loneliness. Henry was able to win over his father by showing an interest in mechanical devices and an eagerness to accompany his father on various trips. One of these took place in 1855, when father and son visited a display of George Scheutz's calculator. The machine was made by a Swedish printer after reading an article on the Difference Engine in the *Edinburgh Review* in 1834. In numerous ways it was considerably different from Babbage's machine. It was a much smaller unit, consisting of four differences and fourteen places of figures, and it could print tables.

Scheutz was employed to publish a technical journal for civil and mechanical engineers. When his son was twenty, they applied to the Swedish Government for funds to work out their machine. Twenty-seven hundred dollars was appropriated, and together they began to work on their model in 1837.

The Scheutz calculator was employed to compute a mortality table—a table of the probabilities of the length of human life. The Scheutzes had very little data to work on, but they had the life statistics of several cities in England, and from them they made the table, which was published in book form and used by insurance companies for many years. It was the first mortality table ever made, and it was accurate.

A description of the original Scheutz machine appeared in the June 30, 1855, issue of *The Illustrated London News:*

This calculating Machine which has created great interest in the scientific world, is the invention of Messrs. George and Edward Scheutz, of Stockholm, and is called by them a Tabulating Machine. It calculates any table not requiring more than four orders of differences to fifteen places of figures, and stereotypes the results to eight places of figures, with proper correction for the last figure, besides five places of figures in the argument. The whole machine is about the size of a small square piano.

The calculating portion of the machine, as appears in the front of the Drawing, consists of a row of fifteen upright steel axes passing down the middles of five rows (fifteen in each row) of silver-coated numbering-rings, each ring being supported by, and turning concentrically on, its own little brass shelf, which has a hole in it sufficiently large to allow the steel axis to pass through without touching. Round the face of each ring are engraved the ordinary numerals, one of which appears in front at a time; and the numbers shown in any horizontal row of rings are read as in ordinary writing. The first row shows the resulting number or answer to fifteen places of figures, eight places of which the machine stereotypes. The second row of rings expresses the first order of Differences, if necessary to fifteen places of figures; and the third, fourth, and fifth row of rings similarly show the second, third, and fourth orders of Differences. Any row can be made to show and calculate with any numbers expressed according to the decimal scale, such as the number 9865432105789; the first eight figures of which if in the top row would be stereotyped. Or (by simply changing two perpendicular rows of rings) it can show and calculate with numbers expressed in the sexial system-degrees, minutes, seconds, and decimals of a second—such as 87 deg. 43 min. 24.687356402 sec., which result, if it appeared in the upper row of rings, would be stereotyped 87 deg. 43 min. 24.69 sec.

The proper argument to each result is also stereotyped at the same time, and in its proper place. Nothing more is required than to set each row of figure-rings to Differences calculated from the proper formula, and place a strip of sheet lead on the slide of the printing apparatus; when, by turning the handle (to do which requires not so much power as can be exercised by a small turnspit dog), the whole table required is calculated and stereotyped in the lead. By stereotyping in the lead is meant that the strip of lead is made into a beautiful stereotype mould, from which any number of sharp stereotype plates can immediately be produced ready for the ordinary printing-presses. At the usual rate of working, 120 lines per hour of arguments and results are calculated and actually stereotyped ready for the press. The Machine which has been brought to England by Messrs. Donkin, has been kindly shown and explained on several occasions to various scientific persons at the rooms of the Royal Society, by Mr. Gravatt, F.R.S.

The article concluded by stating that Mr. Gravatt had explained the machine to Prince Albert of England.

Few people expected Babbage to congratulate Scheutz on his work. Why should he? He had spent a lifetime trying to perfect and build a Difference Engine, and along came Scheutz, who finally succeeded where Charles had failed. But greatness is revealed in many ways. Babbage welcomed the Scheutz innovation. He did everything he could to promote it and was instrumental in its being awarded the French gold medal for Scheutz.

The first model was bought for five thousand dollars in 1856 and sent to the Dudley Observatory in Albany, New York (it is now the property of the Victor Business Machine Company in Chicago), and a duplicate was made for the British Government and used in the Registrar-General's Department.

Babbage was pleased with how his government and others had accepted Scheutz's machine. Why not? It was Babbage's original plans that had inspired Scheutz. Scheutz's engine was much simpler, and therefore its construction was easier. Most important, its acceptance proved that a calculating machine of this type not only could be built but would also be useful.

Old age had now come to Charles, and in 1862, at seventy-one, he agreed to have the completed section of his first Difference Engine shown to the public. Although it was presented at the Great Industrial Exhibition in London, it was placed in a corner far from the major displays. As Babbage said, the Difference Engine was eventually shown in "a small hole, four feet four inches in front by five feet deep." It was surrounded by other exhibits, and no more than six or seven people could glance at it at any one time.

Babbage's long series of disappointments led him to say that he had never had a happy day in his life, and as one of his friends recorded, Babbage spoke "as though he hated mankind in general, Englishmen in particular, and the English government and organ-grinders most of all."

Having authored eighty papers in his life, Babbage decided to write his autobiography, *Passages from the Life of a Philosopher*.

One direct outcome of his autobiography was the attention given to Babbage by his American friends and the United States Government. The U.S. was looking for the best calculating machine available and wanted to buy

Babbage's Engine. Charles replied to this kind invitation that his machine would not satisfy the needs of the agency and that in fact, even if it could, it was not up for sale.

At this time, after nearly forty years on his engines, the machine of Babbage himself began to wear out. "Mr. Babbage is looking wretchedly and has been very unwell. I have done all I could to persuade him to leave town, but in vain. I do fear the machine will be the death of him, for certain I am that the human machine cannot stand that restless energy of mind."

His work was over. On October 14, 1871, Charles became seriously ill, and death was upon him. "It's a long time coming." He turned to a friend, "Now I am going, as they call it, to the other world: ask me any question you like as to my feelings or thought, and I will tell you." He was at ease, and his thoughts did not turn to his youthful dream about returning to tell his friends the hidden secrets of the next world.

The clock struck eleven on the night of October 18, 1871. Charles whispered to his son, "What o'clock is it, Henry?" There was a reply, and within the half-hour, Babbage, two months short of his eightieth birthday, was dead.

People spoke of the death of Charles Babbage as a recurrence of the words of Antony: "I have been sitting longer at life's feast than does me good. I will arise and go."

Even with death, Babbage could not be left alone. There were some who felt that he had been perhaps the most original of all thinkers since Sir Isaac Newton. They believed that what the two men had in common was their extraordinary powers of intellectual introversion. Like Newton, he first pondered and questioned his facts, added light and substance to them by continuous examination and then proceeded to uncover the principles on which the facts depend.

The hope to discover something unique about Babbage's mind was great. Sir Victor Horsley, a surgeon, questioned whether Babbage's great calculating machine was to be found in his brain. After a thorough examination of his brain, it was determined that there was nothing about it that could explain his genius. The brain was preserved by the Royal College of Surgeons of England, where it still can be found.

Charles Babbage was a complex man who never ceased to be on the run. Even with death approaching, he showed an unusual desire for continued life. Although he repeated on his deathbed that he hated life, he did not want to die. He once remarked that he would be happy to give up the remaining years of his life if he would be permitted to live three days five hundred years later so that he could see with his own eyes what new discoveries had been made.

Of course Babbage needn't have requested five hundred years, for were he to have seen his world a mere one hundred years later he would have been astounded by the day's wonders—and he would rightfully have been acclaimed the genius that he was. He would have been flowered with praises, for today's computers can find their basic roots in the "dream of Babbage."

His son's tribute was a dedication to his father's tireless efforts. Henry Provost Babbage spent years accumulating materials about his father. He gave the unfinished Analytical Engine to the Science Museum in London, and the numerous volumes of his father's correspondence went to the British Museum.

Although many had found fault with Babbage during his lifetime, no one could accuse him of being inconsistent with his blinding love for research and precision. Babbage knew his own worth and asserted his rightful claims. Never victorious, he was also never vanquished. Above all, he was heroic in what he said and all that he did.

He was a philosopher of the greatest magnitude:

My engine will count the natural numbers as far as the millionth term. It will then commence a new series, following a different law. This it suddenly abandons and calculates another series by another law. This again is followed by another, and still another. It may go on throughout all time. An observer, seeing a new law coming at certain periods, and going out at others, might find in the mechanism a parallel to the laws of life. That all men die is the result of a vast induction of instances. That one or more men at given times shall be restored to life, may be as much a consequence of the law of existence appointed for man at his creation, as the appearance and reappearance of the isolated cases of apparent exception in the arithmetical machine.

Babbage held to the strongest and loneliest philosophy—that no man could be much helped or hindered by anyone but himself. He both made and destroyed his life's effort. There was not a position that he ever sought that he gained. On the other hand, there was not an invention connected with his name that he could not have perfected had he had sufficient funds at his command.

For Babbage the lasting monument will not be found in a museum. For this genius the realization of his research and experimentation would have certainly been more important. He had shaped and predicted many of the things to come. One day in 1944 the British Magazine *Nature* noted the completion of Harvard University's Mark I computer as "Babbage's Dream Come True." The director of the Mark I project said: "If Babbage had lived seventy-five years later, I would have been out of a job." Charles Babbage had been through it all nearly a hundred years before.

4

George Barnard Grant

The Forgotten American Innovator

George Barnard Grant was one of the major mechanical geniuses of his day. In the designing of machines and the generating of power through the interaction of gears, he had no equal in the United States.

His concern for the state of the art was revealed in prose:

To most mechanics a gear is a gear,

> "A yellow primrose by the shore,
> A yellow primrose was, to him,
> And it was nothing more";

and, in fact, the gear is often a gear and nothing more, sometimes barely that.

For George Grant, the secret of proficient machinery was in the development of superior tooth-action gears. "The present object is practical, to reach and interest the man that makes the thing written of; the machinist or the millwright that makes the gear wheel, or the draftsman or foreman that directs the work, and to teach him not only how to make it, but what it is that he makes."

This importance Grant stressed over and over again. A gear is more than just a piece of metal, it is a mathematical-engineering phenomenon essential to machine operation:

But, if the mechanic will look beyond the tips of his fingers, he will find that it can be something more; that it is one of the most interesting objects in the field of scientific research, and not the simplest one; that it has received the attention of many celebrated mathematicians and engineers; and that the study of its features will not only add to his practical knowledge, but also to his entertainment. There is an element in mathematics, and in its near relative, theoretical mechanics, that possesses an educating and disciplining value beyond any capacity for earning present money. The thinking, inquisitive student of the day is the successful engineer or manufacturer of the future.

Grant was thorough; he didn't miss anything in describing the various forms of gears and their immense value to industry. In his "Treatise on Gear Wheels" he talked of the spur gear, involute gears, cycloidal gears, pin-tooth gears, twisted, spiral and worm gears, irregular and elliptic gears and bevel gears. His importance to American business was considerable. His role in assisting the major industries throughout the country was great. From a satisfied customer in 1891:

Dear Sir:

Two years ago you sent me one of your books on Teeth of Gears, and I have replaced all of the gears in our brick machinery with new ones from your involute odontograph table. I find that we now have the finest cast gears in the world. I do not understand why pattern makers don't catch on to your book. It is a sight to see the gear patterns that are made by some men who are called good pattern makers.

George Barnard Grant was born in Gardiner, Maine, on December 21, 1849, and prepared for college at Bridgton Academy. For three terms after that he studied in the Chandler Scientific School of Dartmouth College, and in 1869 he entered the Lawrence Scientific School of Harvard

College, where he completed the four years' course in three years, receiving the degree of B.S. with the class of 1873.

Only about ten years of his life were spent in the pursuit of a calculating mechanism, but they were to be crucial ones, both for Grant and for those who would follow him in the United States.

While at college he became immensely interested in calculating machines and studied the engines of Babbage and Scheutz. By the time of his graduation he had two patents and had written an article, "On a New Difference Engine," for the August, 1871, issue of the *American Journal of Science and Arts* in which he paid tribute to Babbage and Scheutz:

> The great labor and expense involved in the construction of reliable astronomical and nautical tables by mental computation, as well as the impossibility of getting them entirely correct, suggested to Charles Babbage the idea that this work might be done almost entirely by machinery, and the machine he invented for that purpose has become famous, as one of the most complicated and costly pieces of mechanism every contrived. . . . Babbage's idea was carried out more successfully by Edward Scheutz, and the two machines constructed on his plan are the only ones ever built for this purpose.

Grant's original curiosity about calculating machines came from his school studies before he read about Babbage or Scheutz: "The idea of contriving a machine for calculating tables first occurred to myself while laboriously computing a table for excavation and embankment. Having never heard of either Babbage's or Scheutz's engines, I imagined it an easy matter, but gave it up in disgust after some study."

The year before Grant wrote "On a New Difference Engine," he heard about Babbage's compelling work and, with the energy of a protégé, he submerged himself in developing his own engine. He designed a machine that might possibly have worked, but he was unable to convince anyone that it would do so and abandoned it again.

Several months later his teacher, Professor Wolcott Gibbs, saw his plans and continued to encourage him to design his difference engine and not to feel discouraged. Grant wrote: "Though I have built no large machine, the efficiency of the design for its purpose may be considered as having been proved."

George's engine consisted of two major components, a calculating part and a printing unit:

In the printing part, the calculated results are stamped into a sheet of lead, wax or other plastic substance, from which a stereotype plate is taken for printing the table, thus avoiding constant error in copying the numbers and setting them up in type from manuscript.

The calculating part consists of the main wheels on which the first terms are set up, the additions made, and from which the calculated results are taken by the printing part; the driver makes the addition, and the carrying apparatus.

George Grant's concepts were startling to the audiences that he addressed. Few had even heard of the European achievements of Babbage and Scheutz, and even fewer were willing to accept that an American could outshine them, and especially one so young.

Gradually he became the accepted pioneer in the United States, and before long his name became equated with that of Scheutz and the power of his calculating engine was compared with those of the Swede. "The size of a completed machine would vary with the capacity. An engine of the same capacity as that of Scheutz, would be three feet long, twelve inches high, and eight inches wide. The cost is estimated at from two to three thousand dollars."

George Grant bordered on mathematical genius, and

when working on a problem or a theory it was not un-
common for him to stay at his desk around the clock, for
two or three days at a time, quite often forgetting to eat.
Four years were to be spent in this dedicated fashion perfecting his machine.

In addition to acknowledging the work of Babbage and
Scheutz, Grant gave much of the credit for designing his
engine to Charles Xavier Thomas, who had constructed a
working calculating machine in 1820.

Thomas was a director of an insurance company in
France and patented a device that he called an "arithometer."
"A recording wheel is in gear with a driving wheel having
a variable number of teeth, and the number added to the
recording wheel at each turn of the driver will depend on
the number of teeth exposed." Thomas varied his number
by placing nine rows of teeth side by side, having from one
to nine teeth each, and made the recording wheel movable,
to be placed to gear with either row at pleasure.

This was Grant's grounding in the use of gears to move
wheels. Thomas' lifetime endeavor would yield a fortune
for George Grant in later years. At this point, however,
Grant's sole interest was in developing his own machine,
and he found that Thomas had given him the necessary
spark of innovation and practicality.

The Thomas machine was manufactured in Colmar,
France, and also became known as the Colmar machine.
The device was the first solution of an old and well-studied
problem. For its day it was the finest device. But it was too
complicated a structure, a mass of small cog wheels and a
delicate mechanism, which required close adjustments, fell
easily out of order and required the best skill in repairing
and the greatest care in handling. However, it did work and
was used for many decades—and it can still be found in
service throughout France.

In assessing Thomas' machine, Grant said, "The needs of the present time required, and the state of the art at this day admits of, a better design, one that is simpler and of more substantial construction, easier to put in order and easier to keep in order, and better suited for the use of those but little accustomed to machinery."

Grant must have felt like a child wanting to learn how to ride a bicycle without ever getting the chance to sit on one. He wrote letters and made numerous presentations in hope of finding someone to give him the necessary funds to make it possible for his project to commence. At first he met with closed ears or disbelief. Finally, he was able to find the monies he needed to build his device. Through the contacts of Professor Wolcott Gibbs, Fairman Rogers financed the construction of the machine in time for its exhibition in the 1876 Philadelphia Centennial. Rogers provided most of the ten thousand dollars needed to complete the machine but wrote into the contract one condition —that the machine be donated to the University of Pennsylvania. Another of Grant's machines, used by the Provident Mutual Life Insurance Company for twenty years, was given to the Franklin Institute in Philadelphia.

Grant's Difference Engine was a monstrosity. It weighed two thousand pounds and contained fifteen thousand parts. It was the size of a piano, stood five feet high and was eight feet long. But it worked and worked well. At the Centennial Exhibition Grant displayed other machines as well. There was his "Barrel" or the "Centennial," and his "Rack and Pinion" calculator, which sold 125 copies.

Grant gave us advanced calculators and contributed considerably to the understanding of difference engines, which were to become the foundation of modern-day computers. In addition, he performed another great service in helping to overcome some of the original prejudices against the use of mechanical devices in the business office.

The muscle of his devices was the gear, and like so many of the pioneers in the development of mechanical contrivances, Grant became involved in cutting gears—gears for calculators, gears for all kinds of machinery. He became one of the founders of this industry in the United States and made a fortune from his earlier experiences with difference engines and from his brilliant grasp of the work of Charles Xavier Thomas. As an entrepreneur he founded the Lexington Gear Works, Grant Gear Works, Philadelphia Gear Works and Boston Gear Works, the latter three still in existence.

He was a sensational businessman. With a lightning mind, he was quick to decide and even quicker to make judgments. He had small patience with inefficiency and would not tolerate incomplete information supplied with receiving bids and orders for his gears. Insufficient information, he said, caused "blunders" and "needless correspondence." Regarding drawings, he said, "I am not responsible for errors, even my own, that can be avoided by furnishing a complete specification with working drawings." As to his power of decision making, Grant said, "I am not responsible for my errors of judgment concerning details that are left to my selection. State exactly what you want and you will be suited every time."

George Grant said and wrote what he believed in. To avoid difficulties on unit prices and rush deliveries he was most precise.

PLEASE DON'T: Don't ask a price on one gear and say you are going to have a thousand later on. It is harmless, but it makes me tired.
HURRY JOBS: Don't expect me to hurry for I won't hurry.

I put everything through as quickly as possible with safety and certainty. I know you won't hurry to pay for spoiled work.

The list price is a humbug unless firmly adhered to, and list prices on iron gears are not adhered to. It takes no longer to write for a net price than for the discount on the list price. . . . I have abandoned the common practice of a fixed list and a variable discount and instead have adopted the method of definite net prices. On receipt of a definite list of gears wanted, giving the list numbers of the gears, I will promptly return a definite "net price" at which I will sell that whole lot within one month. I will send separate prices on lots of cut and cast gears, but I will not price the separate gears of a lot. The net price will be lower for a large lot than for a small one, and lower to a customer who sends cash or is known to pay promptly than to one who is likely to defer payment.

His style of appealing to future customers was curious, brave and sometimes arrogant. As an incentive to potential users of his gears, he said: "To a new customer who is not familiar with the quality of my list goods, but not to one who knows what they are, I allow the privilege of immediate return at his own expense. The return must be immediate, and I cannot permit the slightest use or service. Under no circumstances can I permit the return of special gears, as they are wholly unsalable."

Not willing to allow emotions to override objectivity in conducting his business, Grant held firm to his convictions. For those wishing credit and loans he was to follow the strictest of regulations:

Credit is determined by Bradstreet's Mercantile Reports. If that report is good I am willing to run an account; but if there is no report, or if it is not good, I positively will not trust. Terms to a business concern that is well rated in Bradstreet's are thirty days, and I draw at sight to settle an overdue account.

Grant's motto was surely "In God we trust—all others pay cash."

Loans were the common practice of the day, but George wanted to protect himself against unnecessary embarrassment or misunderstanding.

LOANS: If a stranger should call at your office and request a loan of ten dollars for a month he would not be likely to get it. If you order my goods, and your name is not in Bradstreet's, you are a stranger to me. A ledger account is a loan for a month. DRAFTS: Don't get mad because I draw on you. Some concerns like to pay by draft, and others do not. I do not know whether you do or not.

Grant had little confidence in methods of freight delivery of his merchandise:

I decline to send by freight except at customer's risk of damage, loss and delay. Goods wanted in short time must go by express. I charge 25 cents cartage on freight to depot, when package weighs less than 100 lbs. Delivery is complete when goods are delivered to mail, express or freight here. . . . If you are patient and like to sit down and wait for things to turn up, just order a small package sent by freight.

Grant was a wit who had a command of sarcasm in business. His style was the ultimate in entrepreneural showmanship. His company made a hand gear-cutter to permit the user to do his own work at the plant. Naturally, for Grant, this would fall far short of quality, since it would mean that the gear cutting would not be carried out in his own factory. He said therefore:

We cannot answer inquiries as to how many gears it will cut, or how smooth it will cut the teeth, except that it is as good as any machine of its grade in all respects. The product of the jobbing gear cutter, like that of the jack knife, is entirely dependent on the skill and experience of the workman. An active man will do more work than a lazy one, and a badly made and dull cutter will not cut a good gear on this or any machine.

He knew how to attract customers, and he knew that he

was manufacturing the best gears in the world. In the introduction to his book on *Gears,* which sold for sixty cents with paper binding and a dollar with cloth binding, he said:

With every copy sold at retail I send a coupon good in trade for the full amount paid. I am sure this book will please any one in my line of business that I refund the price paid if the book is returned promptly to me, for not one book a month is returned.

The book became a classic in the field and can still be purchased.

Grant's contribution to the world of business and mechanics having been made, he spent his last remaining years in California where he was actively engaged in catching and mounting butterflies. Throughout life he was interested in botany and collected one of the largest private herbariums in the country. In 1918, George Grant died at the age of sixty-nine in Pasadena, amid the quiet splendor of rolling hills far from the pressures and excitement of factory life.

Thus ended the legendary figure in the jobbing-gear business. Although most of his working life was devoted to industry, Grant must be credited as the first American to design and sell a highly organized and sophisticated calculator and Difference Engine, which is now on display at the Smithsonian Institution in Washington, D.C. The foundation that was to make the United States the leader in the computer field is indebted to the foresight and brilliance of George Barnard Grant.

5

Dorr E. Felt
William S. Burroughs

Perfecting the Versatile Calculator

"I realized that for a machine to hold any value to an accountant, it must have greater capacity than the average expert accountant. Now I knew that many accountants could mentally add four columns of figures at a time, so I decided that I must beat that in designing my machine." These were the words of Dorr E. Felt, inventor of the comptometer, in the year 1886.

By the middle of the nineteenth century, the calculator came into its own as a modern tool in the business office. The art of its development occurred in the United States.

In 1850 the U.S. Patent Office issued a patent to D. D. Parmelee for a key-driven adding machine. He was the first to deviate from the established principle of using numerical wheels and in its place used a long ratchet-toothed bar. Parmelee was also the first to use depressible keys in a calculator.

Seven years later, Thomas Hill secured a patent on a

multiple-order key-driven calculating machine. His apparatus received such attention that it was placed on exhibit in the National Museum at Washington. Although it had some ingenious features and was suggestive of devices of thirty years later, it failed to control the action of the mechanism under the tremendously increased speed produced by the use of the depressible key as an actuating means. On a sudden depression of a key there would be a rapid whirl of his numeral wheel, and this whirl could not accurately or quickly be stopped. Perhaps Hill was naïve enough to believe that the operator of his machine mentally could control the wheels against overrotation.

Unquestionably the first machine to control the carried wheel in a key-driven machine was invented by M. Bouchet in 1885. His device was manufactured and sold, but because it lacked capacity it never became popular.

C. G. Spaulding also provided the control for primary and carrying actuation. However, this control locked the higher wheel in such a manner as to prevent the wheel from being operated by an ordinal set of key mechanisms.

In 1884 a youthful machinist conceived an idea from watching the ratchet feed motion, which was indirectly responsible for the final solution of the multiple-order key-driven calculating machine. After months of effort, he made a wooden model, a "Macaroni Box," which was completed in 1885. Although it was a crude machine, Dorr E. Felt had hit upon the mechanical basis for the modern-day calculator. Said Felt:

Watching the planer-feed set me to scheming on ideas for a machine to simplify the hard grind of the bookkeeper in his day's calculation of accounts.

Therefore, I worked on the principle of duplicate denominational orders that could be stretched to any capacity within reason. The plan I finally settled on is displayed in what is generally known as the Macaroni Box model. This crude model was made under rather adverse circumstances.

The construction of such a complicated machine from metal, as I had schemed up, was not within my reach from a monetary standpoint, so I decided to put my ideas into wood.

It was near Thanksgiving Day of 1884, and I decided to use the holiday in the construction of the wooden model. I went to the grocer's and selected a box which seemed to me to be about the right size for the casing. It was a macaroni box. For keys I procured some meat skewers from a hardware store for the key guides and an assortment of elastic bands to be used for springs. When Thanksgiving Day came I got up early and went to work with a few tools, principally a jack knife.

I soon discovered that there were some parts which would require better tools than I had at hand for the purpose, and when night came I found that the model I had expected to construct in a day was a long way from being complete or in working order. I finally had some of the parts made out of metal, and finished the model soon after New Year's day 1885.

Felt, at the age of twenty-four, had made the first operative multiple-order key-driven calculating machine. He started to manufacture his device in the fall of 1886. With only a very limited amount of money, he was forced to make the machines himself. He was able to produce eight finished calculators before September, 1887. Immediately, two of these machines were put into service to train operators.

In September, 1887, Felt left for Washington with one of his eight original machines and exhibited it to General W. S. Rosecrans, then Registrar of the Treasury, and it was put to immediate use. Another model was placed with Dr. Daniel Draper of the New York State Weather Bureau in New York City.

On November 8, 1887, Felt formed a partnership with Robert Tarrant of Chicago. The success of the comptometer was so impressive that until 1902 no other multiple-order key-driven calculating machine was placed on the market in competition.

In the instruction booklet "Comptometer, The Modern Calculator," Felt's genius was disclosed not only in the usability of his device but also in the descriptive rules for using it. His instructions, in detail, were identified for figuring multiplication, subtraction, division, square root, cube root, interest, exchange, discount, English currency, etc.

By the year 1885, Felt realized the potential for his calculating machine in the banking industry and combined his scheme for recording with the mechanism of the machine he was then manufacturing. Because the printing in his comptometer was accomplished by individualized type impression, legibility of recording as well as accurate addition was obtained. But there were two characteristics of this idea which would make the invention unmarketable.

First, the motor had to be wound, and second, there was no provision for printing ciphers.

Shortly thereafter, Felt manufactured his second recorder, which became the first practical recording-adding machine ever sold that would produce legible printed records of items and totals under the variable conditions that have to be met in such a class of recording. Like his first recording machine, this was a visible printer, allowing each figure to be printed as its key was depressed. The paper had to be shifted by a hand lever at the right of the machine. Unlike the former calculator, the operator was no longer required to perform the extra task of winding up a spring to furnish power for the printing; instead the power for the printing was stored by the action of the paper shift-level, and an entirely different printing device was used.

A young man by the name of William S. Burroughs became fascinated with the Felt machine and began to apply Felt's principles to his own inventive activities. In the end

Burroughs produced a type of recording machine that proved to be more acceptable from an operative standpoint than the recorder made by Felt.

Burroughs' first machine, 1888, was designed to record only the final result of a calculation. This was soon followed with a patent for a machine that combined the recording of the numerical items and the recording of the totals in one machine. He had combined in his machine the printing of the totals with the printing of the items.

The inventor of this machine was born into a home of humble parentage on January 28, 1857, in Rochester, New York. His father, an unsuccessful mechanic, moved his family to Lowell, Michigan, when William was very young. Several years later, the family again moved, to Auburn, New York, where William attended public school.

Not wanting to see his son become a mechanic, William's father was able to place him in a local bank, where he learned the basic concepts of accounting and bookkeeping. Five years were to pass in this tedious and boring work during which William became sufficiently frustrated that about half his time was spent trying to guard against mistakes and the other half searching for mistakes that could not be prevented.

A dedicated worker, William on occasion was found adding and checking long lists of numbers throughout the night.

Over a period of time William's health had been deteriorating from the constant strain of work, and his doctor urged that he seek another means of work. He was convinced that his life would be free from the day-to-day frustrations of an accountant. He left the Auburn bank and relocated in St. Louis in 1882, following in his father's footsteps by becoming a mechanic in a local machine shop.

Joseph Boyer, the owner of the shop, had seen Burroughs

once or twice. He watched young William making a collapsible chicken coop—one that could be folded up out of the way when it was not being used. Their long-standing relationship and ultimate friendship was to blossom several years later into the founding of the American Arithmometer Company.

One day a man appeared and asked, "Can you tell me where I can get a helper?" Boyer looked up and said, "I can't say where you can get a man, stranger, but I've got a boy you can use right well. This boy will do you more good than any man I have in the place."

On one of his assignments Burroughs repaired some machines for Thomas Metcalf, a St. Louis dry-goods merchant. They became friends, and one day Burroughs mentioned his idea for an adding machine. Within a few weeks enough money was scraped together—seven hundred dollars—so that William could start to work on his adding unit.

Burroughs often worked through the late hours of the night, developing a machine that would record amounts on paper, add these figures, and carry a progressive total just as fast as the amounts were listed, so that on pressing a key at any time a correct total would be printed instantly. "Accuracy is truth filed to a sharp point," was his slogan. What a relief it would be, he often thought, if a mechanism could be devised that would make such brain-racking labor unnecessary.

He first drew his adding machine plans on paper, but the cold, dank, and damp weather expanded the sheet and his drawings lost their precision. Turning to a more reliable material, William polished sheets of copper that would not stretch or shrink by a fraction of a hair, and he cut his lines with the point of a needle.

Everything was moving along well until his eyes began to feel the pressure of the intensive work. Once again he altered his strategy and drew his adding machine plans on polished zinc chemically blackened. Under this method his lines showed white against a background of black. This turned out to be one of the most accurate ways of drafting plans.

To Burrough's dismay, the initial seven hundred dollars didn't last long enough. Materials were more costly than he had expected, and the day arrived where he was without funds. In the spirit of the dedicated businessman Burroughs went out to solicit more money. But his first machine proved a failure, and his earlier supporters turned him down. Nevertheless, William persisted and started to work with new plans, new materials and new approaches. He built a second model that failed and then a third. Fifty copies of his most recent machine were built, but they were unsuccessful too. Burroughs was able to make his machine work, but the untrained operators failed him. A sudden jerk of a lever would operate the machine too slowly, or quickly, causing a mistake.

Four of the first ten machines were placed in a store run by R. M. Scruggs, one of the financial backers with Metcalf, and another ten were shipped to New York City.

The failure of these first machines drew bitter complaints from William's business partners and customers. Those who had been persuaded to try out the adding device were even more skeptical now, and Burroughs was compelled to call back all fifty machines.

Burroughs couldn't tolerate imperfection, and one by one, he threw his fifty machines out of a second-floor window in St. Louis. Just about everyone who was involved in this project was prepared to give up, but not William. He was determined to succeed, and he locked himself up in his workroom for three days and nights to build his foolproof machine. At last, he perfected a governor mechanism, which guaranteed that his model would perform consistently, no matter how slowly or quickly the operator yanked the lever.

His automatic governor, called a "dash pot," was composed of a metal cup filled with oil in which a plunger performed similarly to that of a piston in the cylinder of an engine. Burroughs set out to build a unit to absorb the shock of pulling the machine's handle, and he finally succeeded.

On January 21, 1886, in the state of Missouri, the American Arithmometer Company was formed. The stock was divided into four equal units—Burroughs receiving one part for his invention, Metcalf and Scruggs their share, W. R. Pye, Sr., a local entrepreneur, the last quarter. Metcalf was chosen president, Burroughs vice-president and Scruggs treasurer.

Burroughs' adding machine was ready for the massive effort of marketing. Initial estimates were made at about eight thousand units, one for each of the operating banks in the United States and Canada. This decision was a good one, but the banks were slow to respond. Before any return had been secured, more than a quarter of a million dollars, a great deal before the turn of the century, had been sunk into the venture.

Burroughs' first adding and listing machine sold for $475. Foreign rights went for $200,000, and the stockholders were overjoyed that the sales did so well. In 1908 all overseas rights were repurchased, and the company from that time on controlled worldwide manufacturing and distribution.

With his dream realized, William S. Burroughs retired to a house in Citronelle, Alabama. He died on September 14, 1898, after a long bout with tuberculosis, and was buried in the St. Louis Bellefontaine Center Cemetery. A simple but beautiful marble shaft marks Burroughs' burial place. On it is inscribed "Under that stately column reposes a man who was noble in poverty, humble in wealth, and great in his benefits to humanity."

The first piece of property owned by the American Arithmometer Company was an 1892 vault used for storage. It cost $132. In 1895 a two-story building was erected in St. Louis, the first floor occupied by Boyer Machine Company, the second by the American Arithmometer Company. That same year the company changed its product's name from Arithmometers and Registering Accountant to Burroughs Adding and Listing Machines to honor its inventor.

With expanding business and opportunity the company was moved to Detroit in 1905, and the name was changed to the Burroughs Adding Machine Company. An article in the September 28, 1908, *Detroit Journal* praises the Burroughs machines that played an important part in tabulating Wayne County election returns. Headlining the article:

ADDING MACHINES HAD LEADING PART

Were Responsible for Accuracy of Journal Totals

Battery of Burroughs Machines Did Noble Work

With World War II the company moved into the computer field—a natural continuity with its previous experience in the calculator area. In 1950 a Sensimatic accounting machine with programmed control panels was introduced. It was the most modern accounting device in twenty-five years and was the forerunner of the electronic accounting computer machines.

The first major Burroughs digital computer was built in 1951 for Wayne University Computation Laboratory.

At this time, with such a variety of products, the board of directors in 1953 changed the name of the company to its present one—The Burroughs Corporation.

The decade of the fifties was to witness a greater and greater involvement into the computer-hardware business. In 1956 the Electro-Data Corporation of Pasadena, California, a manufacturer of high-speed general purpose digital computers, was acquired. An important contract was received from SAGE to construct a ground guidance computer for ballistic missiles for continental air defense.

In 1957 the first ground guidance computer was delivered to the Air Force at Cape Canaveral (now Cape Kennedy) to guide the Atlas missiles. This was the world's first large-scale operational solid-state transistorized computer and is presently to be found in the Smithsonian Institution in Washington, D.C.

Burroughs was building up a reputation in the field. In 1960 the company constructed computers for the Polaris-firing submarines and was awarded a contract for a computer to enable flight controllers of the Federal Aviation Administration to identify approaching aircraft instantly in crowded airport approach patterns.

Burroughs computers have been active in space exploration. In 1962, Burroughs guidance computers directed astronauts Scott Carpenter, John Glenn, and Walter Schirra into space, and in 1963 steered Gordon Cooper into orbit. In 1965, Gemini III's flight with Virgil Grissom and John Young and the four-day Gemini IV flight with McDivitt and White used Burroughs computers. The corporation still continues to play a major role in exploring the world beyond earth.

This is Burroughs' success story. With ideas borrowed from Dorr E. Felt in 1884, and seven hundred dollars, the company reached $8 billion gross sales in 1966, with over $16 billion projected by 1970. The company has paid a lasting tribute to this dedicated, tireless, practical innovator. Although Burroughs devoted a lifetime to manufacturing adding and bookkeeping machines, entry into the computer field was inevitable. William S. Burroughs remains another one of America's pioneering business heroes.

6

Herman Hollerith

With Jacquard's Help—The Development of the Punch-Card System

The boy dreamed too much. It was not proper for a draw-boy weaver to spend his time on the job thinking about the inefficiency of his machine. After all, he was responsible only for pulling certain groups of cords, which were already set for him, so that the warp in the loom would separate in the correct places to produce the desired pattern.

Joseph Jacquard was accused of being lazy and irresponsible. But the truth is that the tedious process of drawing sets of cords in repetitive order failed to stimulate his active mind. He was convinced that much of his labor could be mechanically performed. What the weaver mistook for laziness was actually the beginning of a system by which the harness in the loom could be activated by holes punched in heavy pasteboard cards.

France was the capital of elaborate ornamental weaving, and the city of Lyons was its silk center. Jacquard found that others had investigated the inefficiency of weaving systems and tried to mechanize the process. In 1725, Basile Bouchon designed and built a remarkable device for automatically selecting the cords to be raised on a loom. He employed a continuous belt of punch paper pressed by a hand bar against a row of horizontal wires in order to push forward those that happened to lie opposite the blank spaces. His machine eliminated excessive typing and looping of threads by a draw-boy.

The machine operated by unrolling a portion of perforated paper and pressing it against a series of needles arranged to slide horizontally in a box. Attached to each cord was a needle. Some of the needles would penetrate the paper, causing the thread to move, while those needles that could not pass through did not allow movement. Bouchon's designers made use of his technique to mark off differing weaves.

Although Bouchon might not have been aware of it, in a very crude fashion he introduced the concept of automatic control of machinery by means of a stored program—a concept basic to today's computer operations.

Three years later, in 1728, also at Lyons, Falcon modified Bouchon's device and used several rather than just one row of needles, each row after the first above the previous one. Instead of the band of paper used by Bouchon, Falcon used a chain of cards. By this new arrangement of needles, the punched cards could depress the needles better than a perforated paper roll, which, because of its convex surface, could press against only one row of needles at a time. Falcon's machine merely required that a card be held against a perforated plate, which he then pressed against the needles. The cards therefore could be strung together in a manner simulating a roll of paper.

In 1745, Jacques de Vancanson combined the principles of Bouchon's and Falcon's machines. His apparatus was placed where the pulley box had previously been, and it could be operated from one position, thereby reducing the number of workers needed. Vancanson made the loom completely self-acting; it traveled backward and forward at each stroke and revolved through a small angle controlled by a ratchet.

One of the silk workers in a Lyons factory was Joseph Marie Jacquard. He had been born in Lyons on July 7, 1752. Like his parents, he was employed as a silk worker and went through a series of apprenticeship training programs. First he learned bookbinding, to be followed by type founding and then cutlery.

When he was twenty his father died, leaving him a small house in the village of Cauzon, near Lyons, and a hand loom. In his spare time he began to invent different improvements in the line of weaving, but without any results other than accumulating large debts.

In 1792, when France was fighting for a constitutional form of government, Jacquard joined the revolutionists and after his return home the following year, he and his son assisted in the defense of Lyons against the Army of the Convention, but Jacquard left the battle when his son was killed near him.

Lyons' Council offered him a room, for working on improvements for weaving at the Palace of Fine Arts, with the condition that he should instruct scholars free of charge. During his stay there the Society of Arts in London offered a reward for a machine for making fishing nets. Jacquard succeeding in perfecting one and traveled under protection to Paris, where he showed and explained his machine before the Conservatorium of Arts and Trades (Conservatoire des Arts et Métiers).

On February 2, 1804, Jacquard received three thousand francs, the gold medal from the London Society and an engagement in the Conservatorium of Arts in Paris and was requested to examine one of Vancanson's looms, which was held in storage. From this brief encounter with the work of Vancanson, Jacquard acquired the final necessary ideas for completing his own apparatus that could make draw-loom operations the task of just one worker.

Jacquard returned to Lyons in the year 1804 to take charge of the workhouse. During his stay there he finished his machine. He combined the best parts of the machines of his predecessors and was the first person to obtain an arrangement sufficiently practical to be generally employed.

In 1806 the Jacquard loom was declared public property, and he received a pension of three thousand francs for his work and a royalty for each of his machines, and he was made Chevalier of the Legion of Honor. Napoleon Bonaparte compelled Jacquard at this time to transfer his invention to the city of Lyons, as well as any of his further inventions.

Until 1810, Jacquard had great troubles because his machine was not understood by the weavers. So violent was the opposition made to its introduction that he was compelled to leave Lyons to protect his life. The Conseil des Prudhommes broke up his machines in public places. Jacquard wrote, "The iron was sold for iron, the wood for wood, and I its inventor was delivered up to public ignominy." But after some years the machine proved to be one of great value, and a statue of him now stands on the spot where the model was destroyed in Lyons. By 1812 eleven thousand Jacquard devices had been sold in France alone, and thousands more were in use throughout Europe. Some can still be found today. At the time of Jacquard's death, on August 7, 1834, at the age of eighty-two, and less than twenty years after its public display, more than thirty thousand Jacquard machines were in operation in Lyons.

Charles Babbage, like Hollerith years later, was to apply Jacquard's method in his own work. He described the method of attaching punched cards: "A full box of cards was placed on the input side, and an empty box on the output side. As each card passed over a perforated prism, or cylinder, just before the stroke of the shuttle, it hung down until it was advanced and folded over the card in front."

Almost eighty years after Jacquard introduced his device, Dr. Herman Hollerith sought ways to apply his technique in the counting and tabulating of information for the U.S. Bureau of Census. Contrary to popular belief, Hollerith learned that the Jacquard device was not a loom but merely an apparatus for controlling the loom. By feeding the loom with instructions from punched cards beautifully intricate cloths were made more easily than before. The Jacquard units were of different sizes and descriptions, some using as few as one hundred hooks, others as many as twelve hundred.

In examining Jacquard's machine, Hollerith found that of its ten parts, one was the card unit: The card consisted of a strong, durable pasteboard cut to the exact size of the cylinder. For cutting or preparing the cards to the required size, a table was used with the different sizes of cards indicated on its surface. A sharp steel blade was adjusted to the side of the table. A heavy knife of sufficient length, and containing a second steel blade, was secured to a projecting bolt on the rear end of the table, allowing enough play for the knife to be easily raised and lowered. The blade of the knife was close against the blade fastened to the table, and when pressed down both blades rest close together. On the front side of the table is a long groove in which is a guide, fastened by a bolt and nut. This guide can be set to suit any of the marks on the table, thus regulating the size of the cards to be cut.

Hollerith found that Jacquard had developed a "Dobby Card Punching Machine" for his smaller units (in this operation the whole card was punched at one revolution or stroke), on "Piano Card Stamping Machines" and "Jacquard Pattern Card Machines" for larger ones. Once the cards were prepared they were laced together either by hand or by machine. Hollerith's principal interest was to determine how different patterns, formed by different card punches, could permit varying thread-grasping wire hooks to fall through holes and thereby control certain operations required to reproduce artists' designs.

In the eighty-six years since Jacquard had perfected his idea, no person had been sufficiently inspired to use or improve the punched card machine for other purposes. Hollerith must have pondered this over and over again when he set out to tabulate the 1880 census of fifty million people in the United States. He found that it took 7.5 years to analyze the data from the 1880 census. The compilation of the census had grown into such a monumental task that it became obvious to him that under the then present system more than ten years would be needed to handle the growing data from the 1890 census. With the obsolete system in use, the year 1900 would be upon the bureau and it would still be tabulating the census from the previous decade.

By 1887, Hollerith, after studying the work of Jacquard and his predecessors, developed a crude machine, designed to record, compile and tabulate census data by the use of a punched paper tape. He had the census information recorded on punched cards for the 1890 count. These cards provided neat, accurate and permanent records that were rapidly processed. The result was that it took only two years to tabulate the information on 62 million people instead of 7.5 years for 50 million.

Herman Hollerith was born in Buffalo, New York, on February 29, 1860, the son of a German immigrant couple. Although he had a normal childhood, he possessed a tremendous dislike for spelling. Rather than face the ordeal of a spelling lesson, he once leaped from a second-story window at school and ran all the way home. Letters written later in his life still show his lack of confidence in spelling. In a letter to his wife in 1895 he said, "Today, I was asked to sit for my silouhette. (I think this is the way they spell it.)" And in another lengthy letter written in 1919: "Please make due allowances for errors of spelling, etc. Life is too short to correct type writing."

Ultimately, he was withdrawn from formal schooling and tutored by a minister of the Lutheran Church. This disciplined exposure proved to be of great value for his gifted mind, and Hollerith was graduated from the School of Mines at Columbia University in 1879, at the age of nineteen.

One of his teachers, Professor W. P. Trowbridge, was also a chief special agent for the Census of 1880. The professor asked Herman to work with him on an important investigation of the census. He accepted, was employed in the Census office on October 20, 1879, and prepared a report on "Steam and Water-Power Used in Manufacturers."

When asked where he had first thought of his census machine, Hollerith would reply, "Chicken salad." An unusual answer, but nevertheless one containing a partial truth. A young lady had been watching the way in which Herman consumed chicken salad one day and invited him to her house for dinner to try some of her mother's specialty. The young lady was Miss Billings, and her father was Dr. John Billings, who was in charge of vital statistics at the Census Bureau. At the table Billings turned to Herman and said, "There ought to be a machine for doing the purely mechanical work of tabulating population and similar statistics." We talked the matter over, and I remember his idea was something like a type-distributing machine. He thought of using cards with the description of the individual shown by notches punched in the edge of the card.

From Dr. Billings the germ of an idea came. Hollerith's other prize was Billings' daughter, whom he married in 1890. Together they had six children, three girls and three boys.

Shortly after his meeting with Billings, Hollerith went to his superior, Mr. Leladd, who had charge of the population division of the bureau, and asked to be taken on as his clerk. "After studying the problem I went back to Dr. Billings and said I thought I could work out a solution for the problem and asked him would he go in with me. The Doctor said no he was not interested any further than to see some solution of the problem worked out."

In September, 1882, Herman temporarily took leave of the Census Bureau. General Francis Walker, another census contact and a friend, left Washington to accept the presidency of the Massachusetts Institute of Technology and invited Herman to come to Boston as an instructor in mechanical engineering. Said Hollerith:

While at Boston I made some of my first crude experiments. My idea at that time was to use a strip of paper and punch the record for each individual in a line across the strip. Then I ran the strip over a drum and made contacts through the hole to operate the counters. This you see gave me an ideal automatic feed. The trouble however was that if, for example, you wanted any statistics regarding chinamen you would have to run miles of paper to count a few chinamen.

Hollerith was always willing to acknowledge the source of his ideas. He credits the greater part of his work at M.I.T. to a chance meeting with a railroad conductor. On a train he noticed a conductor hand-punching his tickets, which recorded a basic description of the passengers. Hollerith concluded that this same technique could be applied by punching a card for each individual in the United States to record the proper census statistics.

After a year, not liking teaching, Hollerith went to St. Louis to continue his experimental work. In the summer of 1883 he returned to government employ and was transferred to the Patent Office, from which he resigned on March 31, 1884. No doubt his short exposure with the patent department was to prove useful throughout his career.

He turned his attention to the construction of his tabulating system. Immediately he devoted himself to the problem of devising a machine for counting population statistics and put in his first application for a patent for it on September 23, 1884. Subsequently, three additional patents were issued to him on January 8, 1889. He later accumulated a total of thirty-one data-processing patents.

In recognition of Hollerith's achievements the Committee on Science and the Arts of the Franklin Institute of Philadelphia awarded him the Elliott Cresson Medal, its highest award, for his outstanding invention. The committee, after seeing the system in operation at Washington, said: "They are of the opinion that it is invaluable wherever large numbers of individual facts are to be summed and tabulated. They consider that the inventor is deserving of the greatest commendation for this useful and novel application of electricity, and strongly recommend that he be granted for his invention the highest award in the gift of the Franklin Institute."

In 1890, Hollerith received his Doctorate of Philosophy degree from Columbia's School of Mines under somewhat special conditions: he was granted permission to pursue the required course of study away from the university. For his dissertation he submitted a paper on "The Electric Tabulating System."

Of the many awards that were presented to him, the Paris Exposition Medaille d'Or, and the Bronze Medal from the World's Fair of 1893, were particular prides of Hollerith.

The basic theory underlying Hollerith's machine was to represent the various characteristics of the population or other items required to be counted, by holes punched in cards or strips of paper in specific locations, and then to count the holes in each location through electrical contacts. Hollerith said before the Royal Statistical Society in London on December 4, 1894:

The system of electrical tabulation may perhaps be most readily described as the mechanical equivalent of the well-known method of compiling statistics by means of individual cards, upon which the characteristics are indicated by writing. As it would be difficult to construct a machine to read such written cards, I prepare cards by punching holes in them, the relative positions of such holes describing the individual.

His approach was one of systematic detail. Once the cards were punched, the positioned holes served not only to direct the count to the proper register on the tabulating machine but also to direct the card to the proper pocket in a sorting box operated in conjunction with the tabulating machine. They also simplified hand-sorting, of which there was much to be done prior to the introduction of the electric sorting machine, late in the 1900 census period, since groups of cards with the same punched symbol could be picked out by sighting through the significant hole or through the use of the sorting needle. This advantage was especially manifest when there were large numbers of cards with the same characteristic; for example, white, in the color classification of the population in a northern area.

The "Statistical Engineer," as Hollerith was called, used a continuous roll of paper, rather than a series of cards, in his first tabulating machine. The unit proposed for the 1890 census used manila cards for data. The first such set of cards was made in 1887 for the city of Baltimore. The cards were 3¼ inches deep and 8⅝ inches long, with three rows of thirty-two punch positions across the top of the card and three rows across the bottom. Hollerith had abandoned the continuous strip and took up individual cards: "Some of the very earliest work I did was for the city of Baltimore where I compiled the vital statistics by punching a card for each death with a conductor's punch. I punched down one side across the bottom and then up the other side of the card. The card was considerably larger than the present card."

The card planned for the 1890 census was to be shorter, 6⅝ inches by 3¼ inches, with punch positions occupying the whole surface of the card.

Before the Hollerith system was considered for the census, two more radical changes had to be made. The entire face of the card was divided into twenty-four columns of quarter-inch squares, 288 in all, of which four columns at the left were reserved for geographic identification. And an entirely new machine was designed for punching the cards—two new machines, in fact, one for the 240 spaces comprising the body of the card and one, designated the "gang punch," that punched several cards at one time, for the geographic identification section.

Herman Hollerith worked as an independent inventor, with no committment from the census authorities to the use of his invention. However, he was constantly in touch with Dr. Billings, and Hollerith did plan his system keeping in mind the requirements for tabulation of the population census.

The crucial test was made. In compliance with a suggestion made by the Secretary of the Interior in 1889, the Superintendent of the Census, Robert P. Porter, appointed a committee to consider the method of tabulation to be employed in the Eleventh (1890) Census. The committee recommended testing three systems in four districts in the city of St. Louis to determine which was easiest to transcribe and to tabulate. The results showed that Hollerith's technique required only about 78 hours; the nearest competitor required more than 155 hours. The time Hollerith required for transcribing information onto cards was less than that his competitors required, but by far the greatest advantage was in the tabulation, for which the Hollerith system required only about one-tenth of the time of the competing methods.

These three methods were now put to the test, four enumeration districts of the Census of 1880 in the city of St. Louis being taken. It was found that the time occupied in transcribing their contents by the Hollerith method was 72 hours and 27 minutes; by the Hunt method, 144 hours and 25 minutes; by the Pidgin method, 110 hours and 56 minutes. The time occupied tabulating was found to be as follows: Hollerith's electric counters, 5 hours and 28 minutes; by the Hunt slips, 55 hours and 22 minutes; by the Pidgin chips, 44 hours and 41 minutes. This settled it. The commission also estimated that on a basis of 65,000,000 population, the saving with the Hollerith apparatus would reach nearly $600,000. As a matter of fact, as the saving was based on an estimation of 500 cards punched per day, while 700 is the average, the saving is 40 per cent more than was expected. It is needless to add that Mr. Hollerith's invention was adopted, and that an arrangement was entered into by the Government with its inventor.

The awarding of the 1890 census contract to Hollerith was crucial to his future financial success. "In due time along came the census and it was indeed a brave act on the part of Mr. Porter to award me a contract for the use of the machines in compiling the census. Where would he have been had I failed?"

To the Pratt and Whitney Company, of Hartford, Connecticut, Hollerith wisely entrusted the development and construction of the keyboard punches. The electrical apparatus was from the Western Electric Company.

His tabulating machine was composed of two parts, the press or circuit-closing device and the counters. The press consisted of a hard-rubber bed plate provided with guides or stops against which the punched cards were placed so that the punch position would be always in exactly the same relation with the rest of the machinery. This hard-rubber plate contained holes or cups corresponding in number and position with the 288 quarter-inch squares into which the card was divided for punching. From inside the bottom of each cup a wire extended to a binding post on the back of the supporting framework. Each cup was partly filled with mercury, to facilitate electrical connection, through its wire, with its individual binding post. Above the hard-rubber base plate was a reciprocating pin box, provided with projecting spring-actuated pins, spaced exactly over the centers of the mercury cups. This device with its series of pins was electrically connected as a unit, and, at the time of operation, grounded.

When a card was placed in proper position, against the stops, each punched hole in the card would be exactly over one of the mercury cups; and when the pin box was brought down, most of the pins would be pressed back against their spring, but where the holes appeared in the card, the pins would go through, into the mercury, and thus come into electrical connection with the binding posts on the back of the machine.

A counter was attached in a similar fashion. In order to avoid false counts, or the skipping of a card that was incorrectly punched, the circuits were arranged to ring a bell each time a card was registered on a counter; and the cards that refused to count, and thus did not ring the bell, were laid aside for investigation and correction.

Using Hollerith's machines, the first count of the whole population of the United States as it stood on June 1, 1890, was made:

Practically only six weeks were needed for the gigantic task. The announcements from the Census Office as to population of various sections followed each other rapidly, and a "rough count" for the whole country was ready as early as October 30, 1890. The last returns did not reach Mr. Porter until November 10, but he was able to issue his celebrated bulletin, giving the "official count" of 62,622, 250, exactly a month later, namely on December 12.

At a dinner given by the chiefs of the Census office to celebrate the occasion, Mr. Porter said, "For the first time in the history of the world, the count of the population of a great nation has been made by the aid of electricity" and every single one of our 62 millions "had marched as it were under the vision of the young men and women who had done such remarkable work with such extraordinary rapidity and precision."

Herman Hollerith's machines were largely responsible for this praise. They had performed their job well. For Hollerith his system was rather simple and would often have a standard explanation:

Data cards were punched with holes on spots indicating particular information, such as employment. To obtain the total number of men working in factories, for example, the data cards were placed one by one, over mercury filled cups. The machine dropped rows of telescoping pins onto the surface of the card. Wherever there was a hole, a pin dropped through the hole into the mercury, thus completing an electrical circuit and activating a switch. The closing switch caused a needle on a dial to jump one or more spaces, depending on the information on the card. When all of the cards were passed through the machine, the desired answer showed directly on the dial.

Hollerith soon became convinced of the commercial value of his machine, and in 1896 he organized the Tabulating Machine Company with its first plant in the old Georgetown section of Washington, D.C. All the machinery used in the tabulation of the 1900 census of population was rented from Hollerith.

Toward the end of the 1900 census, Hollerith developed a so-called automatic tabulating machine, into which the cards were fed automatically rather than inserted one by one by hand. When the work on the census of agriculture progressed, Hollerith realized that some more rapid method of sorting was required in order to keep ahead of the tabulating machines, and he manufactured the first modern electric sorting machine.

While the census was underway Hollerith's attention was drawn to the statistics of agriculture:

Here was a question of adding not counting. The only previous work approximating this was in the case of some work I did for the Surgeon General's office. . . . Here was the problem for determining the number of days sick for the soldiers. The reports were made monthly and any case may have been sick from 1 to 31 days, so I had to develop a machine for this and I followed the same lines in the Agricultural statistics. They were certainly wonderful machines. They were operated by weights and I had these weights over against the wall and small wire ropes running from the machines over pullies.

Although these machines were innovative and when truly perfected would prove immensely valuable, the Chief of Agriculture, Dr. L. G. Powers, said that through the use of the Hollerith equipment, the census had cost twice as much as it would have cost done by hand work and adding machines. Powers was not entirely fair to Hollerith. His method might have cost the government more money, but the accuracy and saving of time must be weighed carefully before passing final judgment.

But Herman Hollerith saw the writing on the wall, and although he kept his eye on the Census Bureau, he sought commercial opportunities. His machines were applied to the auditing of freight statistics by the New York Central and Long Island railroads, after which the method was generally adopted by other railroads. When asked by a train official why he had failed to bid for these accounts earlier, Hollerith replied, "I remember distinctly telling him there was one good reason and that was that I did not know the first damned thing about railroad accounts." But he learned and he learned fast.

While busily engaged in his work for various railroads Hollerith went to Russia to introduce that nation to his machines for its census. A large sample of units were made for the Central Statistical Bureau of the Imperial Russian Government in St. Petersburg. Upon his return from Russia Hollerith "was greeted with the news that the machines were thrown out."

Hollerith was to have better success with other foreign countries. Perhaps the most practical demonstration of the value of his system came when his machinery was adopted by the governments of Canada and Austria for their respective censuses.

The Tabulating Machine Company became world famous. A large machine-tool-building company used his tabulating machine for compiling costs, analyzing payroll, keeping track of materials and hence carrying a perpetual inventory. The company was able to receive contracts for analyses of sales and shipments. "We are constantly finding some new item in connection with our business, whereby we can make use of the machine."

A large wholesale house with eight departments carrying 33 classes and 170 subclasses of merchandise made use of his systems for obtaining classified information of sales as to source, salesman, kind of merchandise, salesmen's commissions, individual customer, territory, cost and selling price and other factors that determined a company's gross profit.

A fire insurance company, using the Hollerith system for analysis and classification work, was able to determine amounts at risk, premiums received and losses paid on several hundred classes into which all the insurance written was divided.

There was little doubt that the designs of Hollerith's equipment could service American industry, but Herman was still anticipating receiving the contract for the 1910 census. He was not to have his way.

Shortly after the 1900 census count was made, the bureau decided to set up its own machine shop (later called the Mechanical Laboratory) for the development and building of tabulating machinery to reduce their dependence on renting Hollerith machinery. In 1905-6, Congress appropriated forty thousand dollars for experimental work.

It should not be assumed that the Tabulating Machine Company had given up hope of participating in the tabulation of the 1910 census. Hollerith knew that the Machine Shop had not yet produced an immediately usable automatic tabulator or an integrating tabulator. He proposed, first, to furnish adding tabulators and sorters adequate for the 1910 agricultural census, or if these were not to be needed, by reason of change of method for agriculture, to give the Census Bureau license rights for use of Hollerith-pattern automatic tabulating and sorting machines in tabulating population and vital statistics, these last for the lump sum of $100,000. The bureau declined this offer. Unquestionably, the bureau would have saved money and gained time by accepting Hollerith's offer, but for a variety of reasons it chose not to.

S. N. North, the new director of the Bureau of the Census, was determined to improve Hollerith's techniques and sought better equipment. He selected for the job James Powers, a little-known statistical engineer who had done some work in the processing of masses of data. Powers was clever enough to have the Government agree to allow him to retain his right to patent any machine he developed. North was so pleased with Powers' system that he purchased three hundred punches, related sorters and tabulators for the 1910 census. Shortly after the 1910 census data was collected, Powers left the bureau's service and in 1911 formed the Powers Accounting Machine Company, and the bureau turned to Powers for its major source of equipment. Years later Hollerith was to express his disappointment in losing his influence in the Census Bureau. "I always have regretted that I could not stay in census work long enough to carry out my ideas regarding verification machines."

At this time Hollerith knew that his patents were running out and became aware of the potential competition from Powers:

There is no doubt that these machines are the very best type I ever developed and the only reason I did not continue to use them was that they were rather complicated and especially they would have tied up a lot of capital in building them. The wonder to me is that as these patents have or are about to expire that Powers or some one like that don't take them up. I have heard lots about Powers' sorter but I don't think they would be worth considering along side of a machine built on these old lines.

In all statistical work Hollerith felt that certain stupid

errors would be made despite every effort to avoid them. "I once wrote to the Director of the Census on the subject and after pointing out some of the blunders in his publications I called his attention to the fact that others were guilty in the same way and as an illustration pointed out that in a publication of the city of Paris they carefully specified the ages of three females who died of diseases of the prostrate [sic]." Herman Hollerith's biting wit was his swan song as well.

His last involvement with the bureau came in the famous suit brought by the Tabulating Machine Company in 1910 against Dr. Durand, the Director of the Census, based on the claim that in remodeling the Powers machines the Census Bureau had in effect built new machines that infringed on some of the Hollerith patents. The suit was finally disposed of without significant action. Hollerith never again had any major dealings with the U.S. Bureau of the Census.

In the same year that the Powers Accounting Machine Company was formed, 1911, Hollerith's fifteen-year-old company merged with the International Time Recording Company, the Dayton Scale Company and the Bundy Manufacturing Corporation to form the Computing-Tabulating-Recording Company. The CTR Company was a holding company and in 1924 was renamed the International Business Machine Corporation. In 1933, IBM was reorganized and became an operating corporation.

In 1927 the Powers Accounting Machine Company, through a series of business consolidations, became the Tabulating Machines Division of the Remington-Rand Corporation, which in 1955 merged with Sperry Gyroscope to form the Sperry-Rand Corporation.

To this day, the ghosts of Powers and Hollerith are still competing. Powers had been the applier and Hollerith the innovator. Hollerith was interested primarily in the electrical application of his idea, while Powers saw the commercial value and patented superior mechanical machines. However, when judging the events of history in data processing, Hollerith will stand as the man who was keen enough to see how the accumulation of data and information could be handled and processed more efficiently. What Henry Ford did for manufacturing, Herman Hollerith was to accomplish for data processing—a means for standardization and a format for the interchangability of information.

After he saw his company merged, to later become IBM, Hollerith remained associated with it under a consulting agreement until 1921, and in fact, was awarded his last patent in 1919.

He remained dedicated and interested in improving his efficient machines. In 1923 he wrote of his plans to develop a tabulator, the description of which is remarkably similar to those now in use. However, he became ill and was unable to realize his plans.

On November 17, 1929, in Washington, D.C., a fatal heart attack ended his life at the age of sixty-nine. Had Herman Hollerith permitted his name to be used on his machinery the world would have equated the initial HH with that of the earliest days of modern data processing. But he didn't want it that way. It was the genius of his system that was dominant to Hollerith, and he will be remembered for his many contributions. Machines have become faster and more efficient, but the original ideas of using punched cards as a prime source of input and output in today's modern computers goes back to the nineteenth century when young Hollerith started working on the 1890 countrywide census. Data analysis has never been the same since.

7

Thomas J. Watson

The IBM Empire
and Howard Aiken's Computer

"The trouble with everyone of us is that we don't think enough," said Thomas J. Watson in 1911. "We don't get paid for working with our feet—we get paid for working with our heads." Sitting at a table at a sales meeting, he abruptly wrote on the easel next to him T-H-I-N-K, and the following day he had the word hung in the room with huge letters.

The company was the National Cash Register, and its president, John Patterson, upon seeting the hanging sign, called Watson into his office and instructed him to have the word in every department within the company before the noon whistle blew.

"By THINK," said Watson, "I mean take everything into consideration. I refuse to make the sign more specific. If a man just sees THINK, he'll find out what I mean. We're not interested in a logic course."

This was a portion of the philosophy of Thomas J. Watson, the founder and leader of one of the most remarkable and successful companies in the world. To the general public no organization is more synonymous with the world of computers than is IBM. Few organizations in America have generated a product with such enormous impact as IBM. Still fewer organizations in the United States can boast of an extraordinary ability to continue to market the vast majority of computers sold in this country, and for that matter, throughout the world. IBM dominates the computer business in a way no other giant industrial organization dominates any other major market. General Motors has about 52 per cent of the U.S. auto business; United States Steel has only about 25 per cent of the domestic steel market; Standard Oil of New Jersey has about 15 per cent of free world crude oil production. IBM, by contrast, has installed by dollar value about 75 per cent of all the computer equipment in the United States. International Business Machines is the reflection of one individual—according to Tom Watson Jr., "my father."

The Watson Irish heritage was dominated by change. Even his name was to be changed from Wasson to Watson. The family was called Watson in Scotland and Watson in Ireland; but when a relative settling in Brooklyn married a Catholic, the other members of the family, in wishing to disassociate themselves from the Vatican, adopted the name Wasson.

This couldn't work, of course. Watson's father had accepted the name Wasson for thirty years but could not deny that his family tree was that of Watson. In addition, tattooed to his arm was the name Watson, and so Thomas John Wasson, born on February 17, 1874, was to be renamed Thomas J. Watson.

His early years were spent on an upstate New York farm. His son was to say: "He grew up in an ordinary but happy home where the means, and perhaps the wants, were modest and the moral environment strict. The important values, as he learned them, were to do every job well, to treat all people with dignity and respect, to appear neatly dressed, to be clean and forthright, to be eternally optimistic, and above all, loyal."

His father wanted his only son to be a lawyer, but the young Watson turned this idea down: "I've decided to teach school." But after only one experience as a substitute teacher he told his father, "That settles my teaching career. I can't go into a schoolroom with a bunch of children at nine o'clock in the morning and stay till four."

Watson next chose the world of business. He spent a year at the Miller School of Commerce, twenty miles east of the farm in Elmira, New York, and studied accounting and business subjects.

After his graduation in 1892 his first job was as book-keeper for six dollars a week in a meat market. Although pleased with his salary, Tom said, "I couldn't sit on a high stool and keep books all my life."

A few weeks later he met George Cornwell, a salesman who offered to the community pianos, organs and sewing machines. Cornwell needed an assistant, and Watson was offered the job for ten dollars a week provided he could supply his own horses. This was the beginning of a sales career that was to last for the remainder of Watson's life.

Although he used to say that in the early days he held the horses while Cornwell made the sales, he soon learned some of the Golden Rules of business that he would never forget.

When Cornwell left their company for a better position, Watson was given the sales territory, "and that was the most responsible job I've ever had from that day to this. And I felt more important in it than any position I've ever held, because I was the general manager, sales manager, account-ant, deliveryman—I was the whole organization."

The abacus, probably the most enduring counting system, seems to have originated in the Tigris-Euphrates Valley some 5,000 years ago.

The towering Renaissance figure of fifteenth-century Italy Leonardo da Vinci. Not only was he a great painter but also a genius who could bring an artist's discipline, training and insight to the pursuit of scientific achievement. *(Self-portrait by da Vinci, 1452-1519)*

A drawing of Blaise Pascal (1623-1662), who in his teens invented, built and sold the first calculating machine.

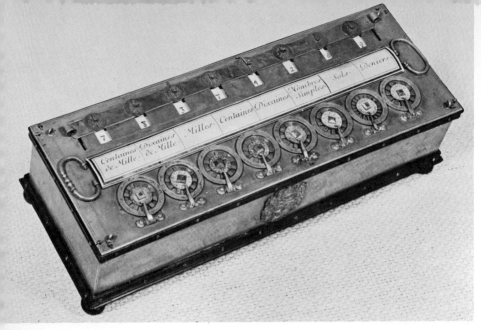

Blaise Pascal started to develop his calculator when he was nineteen years old, and he called it his "pascaline." It was completed in 1643. *(Courtesy of IBM)*

Leonardo da Vinci's theoretical contribution to the world of mechanical calculation remained a mystery until the rediscovery in 1967 in Madrid of two bound volumes of his notebook materials. His calculator model, interpreted from his drawings, showed a mechanism that would maintain a constant ratio of ten to one in each of its 13 digit-registering wheels. *(Courtesy of IBM)*

John Napier, a Scot, developed logarithms in 1616 and invented a device later called "Napier's Bones."

John Napier invented a mechanical
device in 1617 which was arranged of
bone strips on which numbers were
stamped out. "Napier's Bones" when
placed into the proper combination
could perform direct multiplication.
(*Courtesy of IBM*)

A sketch of Gottfried Wilhelm Leibniz
(1646-1716), the builder of an all-purpose calculating machine.

Charles Mahon, the Third Earl of Stanhope (1753-1816), invented the Stanhope Demonstrator in 1777.

Gottfried Wilhelm Leibniz said, "It is unworthy of excellent men to lose hours like slaves in the labor of calculation which could be safely relegated to anyone else if machines were used." In 1694 he built the first all-general-purpose calculating device to add, subtract, multiply, divide and extract square roots. *(Courtesy of IBM)*

A model of the reading dials from the Stanhope Demonstrator. *(Geoffrey Clements, New York)*

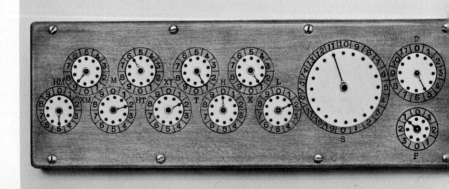

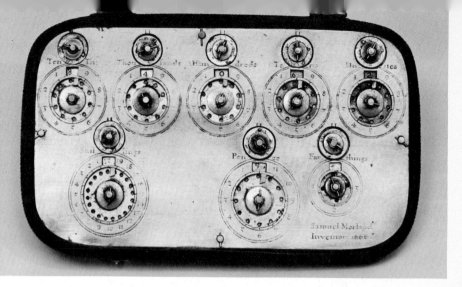

Morland's adding machine invented around 1666. (*Courtesy of IBM*)

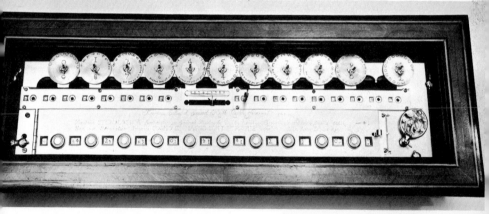

Morland's multiplying machine composed of twelve plates, each showing a different part of the mechanism. (*Courtesy of IBM*)

Sir Samuel Morland (1625-1695) replaced "Napier's Bones" with discs and became the inventor of an operable multiplier. (Geoffrey Clements, New York)

The Stanhope Demonstrator by Charles Mahon was the first arithmetical calculating machine that employed geared wheels. *(Courtesy of IBM)*

Charles Xavier Thomas, a director of an insurance company in France, made a calculating machine in 1820 that is credited with being the first that ever did work practically and usefully. *(Courtesy of IBM)*

The Millionaire was manufactured in Switzerland under the patents of E. Steiger of Germany. A popular commercial calculating machine, it was based on new principles developed by the Frenchman Bollée in 1889. (*Courtesy of IBM*)

In 1885 Dorr Felt designed a multiple-order key-driven calculating machine out of wood. "I went to the grocer's and selected a box which seemed to me about the right size for the casing. It was a macaroni box." Felt called his model the "Macaroni Box." (*Courtesy of IBM*)

In November 1887 Dorr Felt formed a partnership with Robert Tarrant to produce the Comptometer. It was so successful that no other multiple-order key-driven calculating machine was placed on the market in competition until 1902. *(Courtesy of IBM)*

William S. Burroughs, 1857-1898. *(Courtesy of Burroughs)*

A sketch of Charles Babbage shortly after he was appointed Lucasian Professor at Cambridge University, the chair once held by Sir Isaac Newton. *(Geoffrey Clements, New York)*

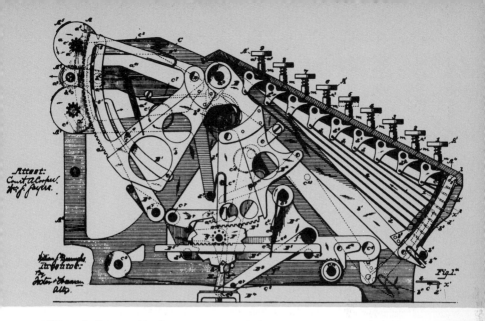

William S. Burrough's first patent was designed to record only the final result of a calculation. This was followed by a machine which combined the recording of the numerical items and the recording of the totals in one machine. *(Courtesy of Burroughs)*

Léon Bollée's machine, invented in 1889 at the age of eighteen, was for multiplication and division. *(Courtesy of IBM)*

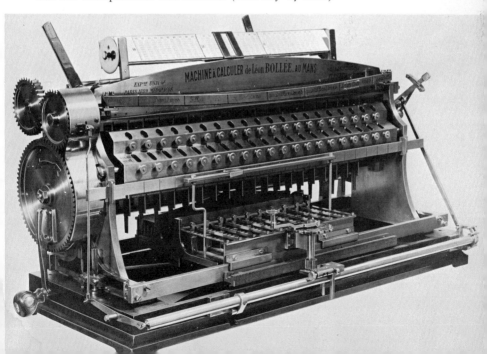

Charles Babbage's Difference Engine was conceived to calculate and print mathematical tables in 1812. Mr. Babbage has been called the "father of the idea behind the modern computer." *(Courtesy of IBM)*

George Scheutz's machine was made by a Swedish printer after reading an article on Babbage's Difference Engine. *(Courtesy of IBM)*

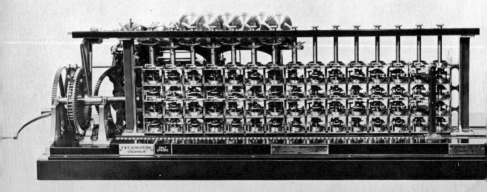

J. Presper Eckert (center, left) making an adjustment on one of ENIAC's electrical boards during World War II. ENIAC weighed thirty tons and occupied more than 15,000 square feet. With its 19,000 vacuum tubes it performed 5,000 calculations each second. (*Courtesy of Sperry Rand*)

The Complex Number Computer frame and teletype console, designed in 1938 by George R. Stibitz at the Bell Telephone Laboratories. (*Courtesy of Dartmouth College*)

BINAC computer followed ENIAC and preceded UNIVAC. It used serial instead of parallel logic. It was the first computer to be programmed internally, the first to use magnetic tape, and it was the first computer to use solid-state elements. (*Courtesy of Sperry Rand*)

UNIVAC I computer—the first all general-purpose commercial computer. It could read information into the system, compute and write information out again simultaneously. (*Courtesy of Sperry Rand*)

A portrait in tapestry of Joseph M. Jacquard.

A sketch of Jacquard loom cards guiding the pattern on a weave of cloth about 1810 in Lyons, France.

An early portrait of Herman Hollerith, who originated the idea of using punched cards as the prime source of data input and output in the 1890 U.S. Census. (*Courtesy of IBM*)

At Harvard University, in 1939, Howard Aiken, with the support of IBM, started working on the Automatic Sequence Controlled Calculator, popularly called Mark I. It was unveiled in 1944 and contained 760,000 parts with switches, components, tubes, wheels and 500 miles of wires. *(Courtesy of IBM)*

Herman Hollerith's tabulating machine was developed for the U.S. Bureau of the Census to assist in the 1890 count. It was the first punch-card machine for data processing. *(Courtesy of IBM)*

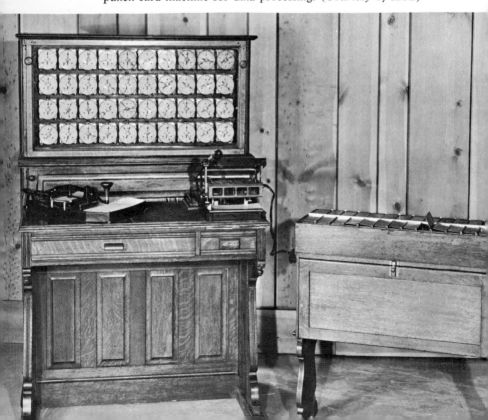

Thomas J. Watson, Sr., founder of International Business Machines, 1874-1956. (*Courtesy of IBM*)

Alan M. Turing (1912-1954) spent a major portion of his life trying to answer the question. Can the machine think? The Automatic Computing Engine (A.C.E.) in Middlesex, England, was built largely from his design (*Courtesy of Mrs. Sara Turing*)

Dr. Corbato worked with Robert Fano at MIT in the 1960's on Project MAC, a massive time-sharing system.

Some early (1920-1930) CTR sorting, key-punch and printing equipment. Thomas J. Watson at this time was president of the Computing-Tabulating-Recording Company, which was to change its name to International Business Machines. (*Courtesy of IBM*)

A view of IBM's 360 third-generation computer. (*Courtesy of IBM*)

By 1894 young Watson realized that a farm town in upstate New York was no place to spend the rest of his life, and he left home for Buffalo. After six weeks of searching for work, he had a series of jobs, first with the Wheeler and Wilcox Company selling sewing machines and later with the Buffalo Building and Loan Association selling stock to the public to finance the company's growth. Also, he borrowed money from his father to open a butcher shop in a residential area of Buffalo. "I had a very sound plan. I was going to keep right on working for this finance company and make money and start more stores until I had perhaps more than Thomas B. Hunter. But the money I had ran out before we got around to starting the second store, so I sold the first one."

At the age of eighteen Watson went out to find his way in the world of business and in three years could show little savings in the bank. Twice he had even consumed all his money; on one occasion he was forced to sleep on a pile of sponges in a store's basement. Philosophically, he said years later, "They say money isn't everything. It isn't everything, but it is a great big something when you are trying to get started in the world and haven't anything. I speak feelingly."

In October, 1895, while arranging a transaction at the Buffalo office of the National Cash Register Company, he applied for a position and after several attempts succeeded in convincing the manager, John Range, that he would be an asset to the company. Range handed over to Watson a pamphlet, "NCR Primer," the first canned sales talk in the country, and said, "You study that and I'll put you on a week from Monday."

In the beginning Tom had difficulty in making a sale, but after several failures, businessmen turned to him and started buying his cash registers. After 3½ years, he was promoted, at the age of twenty-five, to sales agent in the Rochester office. "The reason I was given that territory was that nobody else would take it."

He uprooted his family, and his parents and sisters followed him to Rochester. At times his sister Jennie worked as his assistant, and when she sold two cash registers, he bought her a diamond ring. Tom Watson proved himself a good salesman, and in 1903 he was summoned to the company's main office in Dayton, Ohio, to meet the president, John H. Patterson.

NCR was faced with a problem of competition from their used machines. As the public's interest in the cash register grew, competitors were making a profit selling reconditioned used NCR machines, thus digging into the sales of new registers. Patterson formed a secret subsidiary company to take over all the secondhand business in cash registers throughout the country. He wanted to put other purveyors of used machines out of business, and he chose Watson above his other four hundred salesmen to head the operation.

Upon his return to Rochester, Tom told his father of the opportunity and that he would have the responsibility for spending a million dollars without having to account to the home office how it was used. His father, in his last offering of advice, said, "This is a very fine thing for you. I know you will handle it right. I am going to give you some advice. Before you spend a dollar of your company's money, consider it as your own dollar. Do not spend any of your company's money when you would not spend your own money." That same day he died.

Thomas' new assignment proved to him that he was more than just a good salesman: he was an effective executive. He had to be one step ahead of his competitors. He opened up stores adjacent to his successful opponents, emulated their good points and rejected their bad ones. He met with and hired his competitors' salesmen, undersold on the price of used machines and eventually succeeded in forcing NCR's secondhand-machine competitors out of business.

Within a year after his arrival in the home office in Dayton, Watson became the third most powerful man at NCR. He was now thirty-three.

Sometime later Patterson visited young Watson in his hotel room and after looking over the well-appointed decor said, "I don't think this is the address you ought to have. I'm going to build you a house." He was repaying Watson for his years of dedication. The house was built, and Watson was not allowed to pay any rent. Patterson next bought Watson one of the prestigious Pierce-Arrow cars.

Tom Watson was approaching forty and had spent little time thinking about marriage. Although he was interested in women throughout his younger days, he felt himself too busy to prepare for a life of marriage. He wanted to wait for the right time, when he could find the woman he would always be proud of.

Jeannette Kittredge was twenty-nine when she met Watson. Her father, an Ohioan, was a very successful businessman, at one time president of the Barney and Smith Railroad Car Company of Dayton. In a simple wedding she and Watson were married on April 17, 1913, and they spent their honeymoon in the western part of the country. Patterson presented Watson with a wedding present of a summer house that was especially built near his own country place. Less than six months later Patterson fired Watson.

Although still very much in control of the company, Patterson disliked being looked down upon. He had grown impatiently jealous of Watson's popularity with the employees, especially with the salesmen.

Patterson and Watson's conversations became cool and indifferent. Patterson once said that when he left the National Cash Register Company, no sales manager would ever take over the helm—meaning Thomas Watson.

The final act came at a sales meeting. Patterson had chosen to alter a company policy, and Watson opposed the idea, saying it would reduce sales by 50 per cent. Outraged, Patterson said that the NCR was his company and no one else should be boss. Watson refused to let the disagreement stop here, for his reputation with the company was at stake. Their battle became a company issue. At a board of directors' meeting Patterson was so rude to Watson that Tom threatened to resign. Within two weeks, Patterson fired him.

Leaving his office at NCR for the last time, Watson turned to his friends and said, "I've helped build all but one of those buildings. Now I am going out to build a business bigger than John H. Patterson has." In three months he would be forty years old.

After several offers from such companies as Frigidaire, Montgomery Ward, Remington Arms and a boat company in Connecticut, he arrived in New York to consider taking over the management of a faltering company known as the Tabulating Machine Company, an organization that produced a counting device originally designed for handling the census statistics.

The company's name was changed to the Computing-Tabulating-Recording Company, or CTR, as it was more popularly called. At a board of directors' meeting, Watson explained why he was interested in a firm like CTR. He felt that he would finally have the chance to become a true entrepreneur without capital and accepted a yearly salary of $25,000 and an option on 1,220 shares of stock.

After three months with the company Watson became its president and concluded an agreement between his own organization and its competitor, the Powers Accounting Company. Herman Hollerith, the inventor of the tabulating machine, was interested only in the electrical application of his idea and failed to patent the mechanical verson. James Powers, Hollerith's recent foreman, formed his own company and started building machines that soon became more efficient than Hollerith's. They automatically printed results as compared with hand-posting required by the Hollerith system, an electrical punch instead of Hollerith's hand-operated one, and a horizontal sorter to contrast with Hollerith's vertical sorter.

Six years after Watson had joined the company, gross income had more than tripled, rising from $4 million to almost $14 million. In the year 1920 CTR was doing more bsuiness than it had in its four previous years combined. CTR was also engaged in manufacturing time recorders and scales, but the big profits came from its tabulating machines.

The recession of 1921 brought further expansion plans to a halt. The company newspaper headlined: "No! Hard times are not coming—just soft times going." At the main tabulating construction plant in Endicott, New York, production slowed down and salaries were cut by 10 per cent, including Watson's. Instead of the anticipated sales of $15 million that year, less than $3.5 resulted.

Faced with near disaster, Watson went to Alexander H. Hemphill, president of the Guaranty Trust Company, for financial assistance. Watson explained to Hemphill that he was arranging to have his notes covered, but Hemphill misunderstood and asked him how much more capital he wanted.

Watson replied, "No more, I have been borowing to expand the business, and there is no time to continue that program."

The bank president replied, "I belive so much in the future of your business and in you as the head of it that Guaranty Trust will go along with you, and don't worry about it." The bank had saved Watson's company from collapse.

During the recession his income was cut by 50 per cent, but with a new period of growth on the horizon, he was awarded a new five-year contract, raising his salary to sixty thousand dollars a year, with a broader-based profit-sharing plan.

In his ten years Watson had done a lot with CTR. In the laboratory his researchers developed an advanced tabulating printer, an efficient key punch machine and the more practical horizontal sorter. In one decade CTR had grown by a factor of three, with sales up 300 per cent. Stockholders received dividends three times greater than in 1913, and the market value of the stock was more than five times what it had earlier been, now almost $18 million.

Nineteen twenty-four was an important year in the career of Thomas Watson. The name of the company was changed from CTR to International Business Machines. Watson chose International to describe his high hopes of global enterprise, Business Machines to indicate the diversity of products.

In the same year, Watson was to capture the position as chief executive officer of the company and now was able to dictate all policies.

One of Tom Watsor's most unusual talents was illustrated in the way in which he chose to promote his men. In the spirit of building morale among his workers he occasionally would make on-the-spot promotions of his sales force in some rather unorthodox way. One of his employees was promoted to an executive position because Watson admired his father, another because Watson was charmed by an employee's wife. These pleasures were part of his philosophy:

"The success of every major executive depends on the men under him. Really successful men are pushed up, not pulled up."

Watson realized that the future success of his business would require listening to specialists who were more technically oriented than he was. He learned the value of using experts. "Every businessman must have someone with ability and judgment whom he respects and use him as a confidential consultant and advisor." But he was always cognizant of his role with them: "You men are responsible as individuals for different things in this business, and I am responsible for all of it."

Watson placed emphasis on loyalty, unity, idealism and enthusaism: "You cannot be a success in any business without believing that it is the greatest business in the world. You have to put your heart in the business and the business in your heart." When men joined his company, Watson expected absolute loyalty from them. This kind of intense loyalty, then, became the wellspring of the IBM spirit, the "family spirit," as it was called. As his enthusiasm grew for the family spirit in IBM, Watson drew a sharp line between fellowship and familiarity: "All of us must remember never to let personal friendship guide us in making decisions. Not only can it harm the man you are trying to help but it can harm you."

Watson had a violent temper. He knew how to tear a man apart and sew him up again, as people used to say. But the truth was that a man could never be put together quite the same. If you weren't worth putting together, Watson didn't bother to tear you apart. He was harder on the industrious than on the lazy, harsher to favorites and the powerful than the obscure, but above all he was unpredictable. Within twenty-four hours after attacking someone, he would be filled with remorse and go to great lengths to make amends, sending the employee on a vacation trip or giving him a raise.

During the depression years of the thirties, when most businessmen were struggling to keep their heads above water, Watson was reaching the zenith of his career. His optimism about the future of his company was towering: "Living in the United States is like going to sea in a huge raft. Our feet are always wet, but we never sink."

Even though the office equipment industry throughout the country suffered a 50 per cent decline or more during the depression, Watson was able to hold his own. He continued his building program and instead of mass layoffs, he continued to make equipment and store it for future use. Accumulating an enormous surplus of wares was the great decision of a genius and was to prove to be the most important step that Watson made in making IBM a wealthy corporation.

In 1935, IBM was selected in competitive bidding to handle the bookkeeping of the newly organized Social Security System, one of the largest governmental operations.

IBM's income from their electrical tabulating machines doubled during the years before the Second World War, and new contracts were awarded the company by various agencies of Roosevelt's New Deal. "I think we are coming out of it all right," said Watson when the company found it was doing more business than any other office equipment organization.

Watson ushered in a new philosophy in the business community. He was to instruct his sales force to be interested in selling results, not just machines. He emphasized the *why*, not the *how* of his business; the applications, not the hardware.

To his salesmen he said:

Individuality is the biggest asset that any man has, from the office boy right up to the president of the company. You cannot sell by formula. If it is natural for you to talk loud and once in a while pound the tables, do so. Then you are effective. Read and study, but mainly be natural. You cannot sell by sitting in the office. The only time that pays off is time before a prospect.

Many of Watson's friends told him that he was out of his mind, that businesses were not run that way, that he was certain to ruin his corporation. But Watson held firm and was to show over the years how wise he was.

Work to Watson was the principle means of expression. As *Fortune* magazine commented: "Business with a capital B, straight-forward American business, thrilled him."

It was during the Depression that Watson found himself involved in federal activities. His first major responsibility was the defense of business when it faced criticism: "The present controversy between our government and business is undignified, unfair and destructive and is discouraging to our people who are trying to do right. I believe business should be well regulated. I also believe business should be well treated."

President Roosevelt replied:

Dear Tom:
You are right about the restoration of political harmony and confidence in one another. My middle name is "meet them halfway." Frankly, one of my principal difficulties in the past has been that I have been asked to meet the other fellow ninety-nine per cent of the way. Harmony and confidence are largely a matter of mutuality.

One of Watson's most eloquent roles was to sell the New Deal to the community of businessmen. All his life, Watson was a Democrat (which surprised many of his fellow entrepreneurs), and when Roosevelt was elected, Watson said of him, "People realized that they were listening to a man who had a new idea, and a man who was not afraid to stand before them and say, 'This is the truth as I see it,' regardless of precedent—because precedent has proved itself to be wrong."

Watson discussed the New Deal in every speech he gave during Roosevelt's Administration. For his support the President offered Watson the ambassadorship to Great Britain and the office of Secretary of the Department of Commerce. He declined both, but before the outbreak of the war served as American Commissioner General to the Paris International Exposition.

His role in the Roosevelt Administration combined with his activities as representative overseas encouraged Watson to become more internationally oriented just before the war began. He started to turn his attention to building up a European organization.

"World Peace through World Trade," became the slogan of IBM. His speech making as president of the International Chamber of Commerce throughout Europe made his own name so well known that many of his foreign subsidiaries were named Watson Business Machines.

Watson's role as American representative on the Continent entitled him to meet with royalty on numerous occasions. In the first several months as president of the Chamber of Commerce, he met with six kings and two princes and would for years to come play the role as Roosevelt's unofficial ambassador of good will. About entertaining dignitaries, Roosevelt once remarked, "I take care of them in Washington. I have learned to rely confidently on Tom Watson to take care of them in New York." Watson knew that his contacts with royalty would bring him the necessary contacts to build his world business.

After the start of World War II, Watson repeatedly stated his confidence about the United States future: "America is more nearly self-sufficient within its own borders than any other country" and the country was "too sturdy and resourceful a nation to be bogged down by any contingency created by foreign events."

When hostilities began in Europe, Watson put all the facilities of the company at the disposal of the Government. More than five thousand IBM accounting machines were used in Washington. IBM machines were installed in mobile units that followed United States troops to foreign theaters of operation. A plant was built to supply the government with major ordnance items ranging from bomb sights to airplane engine parts. Watson was awarded the Medal of Merit in recognition of his company's achievement. IBM equipment kept toll of the conflict—made bomb surveys, counted the missing, the prisoners, relief materials and displaced persons. The company machines helped break the Japanese code before the Battle of Midway and predicted the June weather over the English Channel. Watson voluntarily limited the net profit on government war contracts to less than 1½ per cent, setting aside that amount as a fund for the widows and orphans of IBM war casualties. Other IBM war efforts included the production of naval and aircraft fire-control instruments, Browning automatic rifles, director and prediction units for 90-mm antiaircraft guns, bomb sights and impellers and parts for aircraft engines. In addition, thousands of IBM accounting machines handled wartime paperwork in Washington, D.C.

It was time for IBM to enter the computer business. One of the most important persons to bring the company into a leadership position was Howard Aiken. Back in 1937, Aiken, a doctoral student in physics at Harvard University, was attempting to build a machine that would solve polynominal problems. He concluded after two years of research that basically the logic of all digital machines was alike and that a general-purpose computer to work on all types of problems was feasible.

Following a series of meetings with IBM in 1939, Aiken was able to receive a carte blanche from the corporation. In fact, it put at his disposal a team of four top-flight engineers. The result of their five years of work came in 1944 when the Automatic Sequence Controlled Calculator—the Mark I— was unveiled.

The Harvard-IBM Mark I was an enormous automatic electromechanical machine containing more than 760,000 parts, with switches, components, tubes, wheels and 500 miles of wire.

Mark I was put to use at Harvard and had an input fed by punched paper tape, a memory unit, an arithmetic device and an output-reading device. Compared to today's standards it was sloppy and slow, but it was the beginning for IBM. Mark I took one-third of a second to add or subtract, five seconds to multiply and sixteen seconds to divide. One and one-half minutes were needed to determine a logarithm to twenty decimal places.

On campus, Mark I found a home in a red brick physics building, and "when it was working, one could go in and listen to the gentle clicking of the relays, which sounded like a roomful of ladies knitting."

Although Mark I was made obsolete by the twelve-times-faster Mark II, and subsequent models, Aiken hailed his feat as "Babbage's dream come true." In fact, the Mark I operating manual began with a quotation by Charles Babbage: "If, unwarned by my example, any man shall succeed in constructing an engine embodying in itself the whole of the executive department of mathematical analysis . . . I have no fear of leaving my reputation in his charge, for he alone will be able fully to appreciate the nature of my efforts and the value of their results."

There have been some who claimed that Thomas Watson, Sr., opposed IBM's entering the computer field. This was not so. He had been a strong advocate for developing some of the pioneering computers, such as the Mark I and the Selective Sequence Electronic Calculator. According to his son, Thomas J. Watson, Jr., "Some confusion may have developed from the fact that at one stage some of us were urging much larger R & D expenditures in the electronics area. My father was at first reluctant to go along with this; in fact he thought this suggestion reflected discredit upon certain of our key engineers. . . . Actually, my father fully participated in the decision to move into computers. Once the decision was made, he backed our moving ahead into new areas and the borrowing of whatever monies were necessary to advance those ends as enthusiastically as he always embraced new approaches and new ideas."

In 1948, IBM introduced their first commerical electronic computer—the Selective Sequence Electronic Calculator (SSEC). Watson had committed himself and his organization to a new and developing field. The Korean War and competition, primarily from Remington-Rand's UNIVAC, led IBM to produce its model No. 701, a scientific computer twenty-five times faster than the SSEC, one-quarter as large and a computer that could readily be mass-produced for commerical purposes. Within a short period of time IBM's No. 702, No. 704 and No. 705 were so well received that competition began to dwindle. Computers and related hardware equipment soon became a major source of income for the company, and IBM was well on its way to becoming the leader within the field. A new era had come to the company.

Watson had great disdain for communism. Some years after the cold war began, Andrei Gromyko was at one of Watson's parties for UN representatives. Someone asked the

Russian delegate if IBM machines were being used in Russia. "I am afraid not," replied Gromyko. "Your State Department won't let us have any." Watson, overhearing this comment, turned and said, "Don't blame the State Department, I won't let you have any."

Gradually, the future of the corporation was shifting away from Tom Watson. In 1954, when computer sales were going well, a gala party was held to celebrate Watson's fortieth anniversary with IBM. His son, Thomas Watson, Jr., who had become president in 1952, told the corporation's stockholders:

I think the story can be summed up in six significant figures—

In 1914 the gross income of this company was $4.1 million. In 1953 it was $409.9 million.

In 1914 the net before taxes was $489,000. In 1953 the net before taxes was $92.3 millions.

In 1914 cash dividends to stockholders were zero. In 1953 they were $12.7 millions.

As old man Watson explained:

A purchase of 100 shares in 1914 would have cost $2,750. In exercising rights through 1925 the cash investment would have increased to a total of $6,364 for 153 shares. This would now (1954) amount to 3,114 shares, with a market value of $1,029,177, which, with cash dividends of $117,356 paid during this period, totals $1,206,533, compared with the original investment of $6,364.

A few months before his last illness, and well into his eighty-second year, Watson remained active in the company. He traveled to Florida to participate in an engineering conference to plan future IBM strategy. Just two months prior to his death he attended his stockholders' meeting and presented for the last time the annual report.

On May 8, 1956, several days after his forty-second anniversary with IBM, Watson turned over the position of chief executive of the company to his son, Thomas Watson,

Jr., and on June 19, 1956, a fatal heart attack ended the life of Thomas J. Watson.

Twelve hundred persons attended his funeral. His minister said, "Integrity was the root of his character." *Time* magazine referred to him as "one of the first of a new breed of US businessmen who realized that their social responsibilities ran far beyond their own companies," and the New York *Herald Tribune* said Watson was "a man who could well stand as a symbol of the free enterprise system." President Eisenhower referred to Watson as "an industrialist who was first of all a great citizen and a great humanitarian."

The corporation was still in good hands. Throughout his youth Thomas Watson, Jr., was often exposed to IBM. Many times he accompanied his father on business trips. When Thomas Watson, Jr., joined the company, he was assigned to the Wall Street sales territory, where no IBM salesman had ever achieved quota. He chalked up 231 per cent of quota. "That was the only right way," his father once remarked. "He had to make his own records. Otherwise people might feel that he had to have some special help." Watson, Jr., was responsible for starting a program to upgrade and modernize the design of everything connected with IBM, from its memo pads to its huge data-processing machines and its headquarters. He was responsible for introducing color typewriters. The change of command was most appropriate at this time. IBM gross income exceeded the billion-dollar level. There was a fast-moving change in both software and hardware throughout the industry.

In 1967 IBM had more than 220,000 employees in the United States and abroad. The gross revenue from sales, services and rentals was $5,345,290,993; net earnings $651,499,588, a $125,369,366 increase over the prior year.

In appearance IBM today is quite different from the company it was many years ago. Products have changed. The starched collar is gone along with company songs. However, the ceremony of awards—the Hundred Percent Club, the Quarter Center Club, suggestion awards, family dinners, country clubs for employees—still remain. In its attitude, its outlook, its spirit, its drive, IBM is still very much the same company it has always been; the beliefs still remain the same.

The senior Watson started with $4 million in sales and 1,346 employees. An investment of $2,700 in 1914 would have bought 100 shares. Through stock splits and added dividends the original 100 would now have multiplied to 59,320 shares worth $17,825,660. In 1914, when Thomas Watson joined the CTR company, its products were butcher scales, meat slicers, coffee grinders, time clocks and various types of crude punched-card tabulating machines. Today IBM pursues developing new methods, designing new products. The basic drive, spirit and philosophy of the company is still symbolized by the senior Watson, who started the greatest computer empire in the world.

8

John W. Mauchly
J. Presper Eckert
John Von Neumann

ENIAC, EDVAC, BINAC, UNIVAC and Sperry Rand

In 1961 a group of renowned scientists and inventors gathered at the University of Pennsylvania for a commemorative banquet celebrating the introduction of the first true electronic computer, built some fifteen years earlier.

Sitting on the dais was John W. Mauchly, amateur meteorologist, physicist and onetime assistant professor of electrical engineering at the university's Moore School of Electrical Engineering, and J. Presper Eckert, Jr., his former research colleague.

John W. Mauchly was born in Cincinnati, Ohio, on August 30, 1907. After attending Washington, D.C., public schools, he went to the Johns Hopkins University School of Engineering in 1925 as a recipient of a Maryland State scholarship. Gradually his interests shifted, and two years later he transferred to the physics department, which awarded him his doctorate in 1932.

Following the traditional path, he taught physics in several colleges. A number of summers were spent at the Department of Terrestrial Magnetism of the Carnegie Institution of Washington, where his interests turned to the statistical analysis of many forms of geophysical data—atmospheric-electric, terrestrial-magnetic, ionospheric and meteorological.

One of Mauchly's major frustrations was the inability to cope with vast numbers of calculations. He was convinced that more rapid and automatic means for carrying out voluminous numerical work could be evolved. He became completely engrossed in theorizing how electronic devices for computing at high speeds could be developed.

"I did fool with American Can adding machines in my basement when I was a high school kid. . . . Before I got to the Moore School, I'll have to confess that I built an anolog computer, which is not widely known, a harmonic analyzer. I also invented a digital computer, but a very special kind for ciphers . . . the first place I know of where memory or storage features of neon tubes or gas tubes was used for a digital computer. . . . Those were the '30's, the early days. . . ."

In his earlier work, Mauchly spent considerable time with a teletype terminal to determine how efficiently it could be used in the handling of mass data: "I was putting problems through it thinking this was wonderful. I still thought it could be speeded up with vacuum tubes. Another man I didn't know, sort of small and interesting fellow, kept telling me computers were great and this was the beginning of something that would turn out to be revolutionary. 'Don't you think I'm right?' said Norbert Weiner."

Mauchly accepted the challenge and arrived at the Moore School in 1941 with the hope that the large "differential analyzer" it had developed would show how electronics would be used in constructing a device for making rapid calculations.

Mauchly was thirty-four years of age when he met the competent and often brilliant twenty-two-year-old Eckert. J. Presper Eckert, Jr., was born in Philadelphia in 1919 and attended the William Penn Charter School in Germantown. After graduation in 1937 he entered the Moore School and in 1941 went on to teach there after receiving his bachelor's degree. His first course was ESMDT, Engineering, Science and Management Defense Training.

His interests were varied. In addition to teaching he became engaged in designing and building a device for measuring the concentration of napthalene vapor by means of ultraviolet light. Later, he worked on the development of instruments to measure the fatigue limits in metals.

At the outset of World War II, Eckert undertook the designing and building of a device for measuring the strength of very small magnetic fields—a device that was employed in experimental methods for setting off enemy magnetic sea mines and for studying schemes for detection of enemy submarines. After successfully completing this device, he worked on the solving of a number of problems involving radar and on various devices for measuring targets to an accuracy of 2 yards out of 100,000 yards.

In June, 1943, after receiving his master's degree in electrical engineering from the Moore School, he accepted an invitation to work with Dr. Mauchly. Together, they spent the first year of World War II tossing about theory and application with each other. As the war crisis became acute in 1942, the Ballistic Research Laboratory of the U.S. Army Ordnance Department was assigned the task to develop firing and bombing tables for existing and proposed gun-projectile combinations, rockets, missiles and other strategic arms. In order to compute a single trajectory for a given set of conditions, military specialists would normally have to spend nearly seven hours at a desk calculator. Hundreds of operators were needed round the clock to construct appropriate ballistic tables.

An Army ordnance mathematician, then at the Moore School, heard of Dr. Mauchly's ideas, and the commanding officer and the young professor agreed that this was a problem that could be tackled effectively.

The first correspondence was dated August, 1942, with a detailed description of the project. The memo lay buried for nearly a year before it was resurrected from a file by Lieutenant Herman Goldstine, an assistant professor of mathematics at the University of Michigan, before joining the U.S. Army's Ballistics Research Laboratory in Maryland.

Goldstine, along with Colonel Paul N. Gillon, also of the Ordnance Department, was responsible for breaking the bottleneck that then threatened the computation of the firing and bombing tables so urgently needed by the military forces. While attempting to hire some girls in Philadelphia to train as calculator operators, he stole some time and visited friends at the Moore School. Once there, he saw the August, 1942, letter and had it reconstructed from a secretary's shorthand. The project was sent down to Washington for approval.

An emergency meeting was held. In attendance was Professor Oswald Veblen of Princeton's Institute of Advanced Study. After sitting in judgment, Veblen turned to the director of the ballistics laboratory and said, "Simon, back that thing." The federal agency did and requested that Mauchly and Eckert spare no expense or time in coming up with the appropriate tables. Quickly, a team of two hundred people was organized and an emergency crash program started to build an electronic high-speed computer.

Mauchly had suggested the possibility of the project during the summer of 1942, and finally, a year later, a contract for $400,000 was signed with the University of Pennsylvania. Mauchly directed the project, Eckert served as his right-hand man and chief engineer, and Captain Goldstine was appointed to maintain technical liaison with the Ballistic Research Laboratory.

The engineers had few previous success stories to copy. They knew that Dr. Vannevar Bush had built in 1930 at the Massachusetts Institute of Technology, the first general-purpose computer, but this was an analog device good for making judgments about varying inputs and outputs but unable to act as a fast arithmetic calculator.

Bush suggested a subject for master's thesis to one of his students, Claude E. Shannon. The Master of Science thesis was later published under the title *A Symbolic Analysis of Relay and Switching Circuits*. Explaining how relay and switching circuits would be expressed by mathematical equations, Shannon showed the calculus for manipulating these equations to be "isomorphic with the propositional calculus of symbolic logica."

Shannon's thesis laid the groundwork for the use of a binary number system to replace a decimal system in the computer. While working at the Bell Telephone Laboratories in 1937 he was about to demonstrate the parallel between switching circuits and the algebra of logic. True or false values were therefore analogous with open and closed states of electrical circuits. Shannon had defined a universal measure and a universal unit of information: the bit (binary digit). "A bit is the choice between plus or minus; it is the amount of information needed to remove the uncertainty between yes or no."

In another part of the Bell Telephone Laboratories worked George R. Stibitz, a mathematician who joined the "lab" in 1930 after receiving his Ph.D. from Cornell University. His design of the Complex Number Computer, the first program-controlled electrical digital computer in 1938, was a direct result of an assignment to investigate the magnetic circuits of certain telephone relays. "The 'lab's' junk-pile supplied several relays, which I took home to play with. Some dry cells provided power, and I soldered up a few different circuits; one of these was the equivalent of the 'two-way' light switch. I noticed that this circuit was equivalent to one digit of a binary adder, and so I drew and wired a circuit for the 'carry' digit. Then I had a binary adder, to which I added two keys made of tobacco tins, and a pair of flashlight bulbs as output."

Stibitz did a great deal of his pioneering work at home, completing his "adder" on the kitchen table in November of 1937. "I took it to the 'Labs' and pointed out to various colleagues that this device was the basis for an adder to handle any number of binary digits, and that it would be possible to make a relay device that would do the work of a desk calculator."

While his Complex Number Computer was under construction, he began to think about extensions of the idea to more general calculations. "My next proposal was a keyboard-operated computer, in which algebraic operations would be directed from a console, results stored, and others printed out. There was no idea yet of a stored program. I drew up a logical diagram of this device, but it was never carried further."

In September, 1940, George Stibitz delivered a paper at a meeting of the Mathematical Society of Dartmouth College. They rented a "line" and demonstrated the use of the computer by remote control. It was a relatively simple task, since the Complex Number Computer was already working over lines between rooms in the Bell Telephone Laboratories. The demonstration was a huge success. Many electrical specialists were in the audience. According to Stibitz, "Norbert Wiener got his first introduction to the idea of a digital computer at this meeting, as did Mauchly; both of them tried out the CC on the teletype at Dartmouth."

The Complex Number Computer was fast. It could do a nine-digit complex division in thirty to forty seconds. Because of relay-timing uncertainties, Stibitz was forced to slow down the speed. To overcome initial shortcomings, he worked out an error-detecting code. "This code was the bi-quinary one, in which a certain number of relays in each group had to be actuated to represent a number, and if exactly this number was not actuated, the next step was not taken." This first error-detecting system was introduced in his next computer—the Relay Interpolator in 1942, which was ordered by the National Defense Research Council, a pre-Office of Scientific Research and Development Organization.

The Relay Interpolator was designed to operate unattended. Frequently, while showing a visitor around, the guest was asked to plug up a relay with a toothpick and note that the computer stopped when this relay was needed, but would go on to a correct answer as soon as the toothpick was removed. Once Stibitz was embarrassed when asked to diagnose the trouble and tell where the toothpick was, without looking. "Fortunately it was at a point in the circuit which was easily identified from the stage of the problem the computer had reached and printed."

Dr. Aiken's work of 1939 was of the digital nature, but it was an electromechanical device and consequently subject to failure and breakdown through wear and exposure to the elements. Since it functioned with moving parts it had speed limitations also.

Mauchly, Eckert and their team were the first to overcome these disadvantages. Thirty months to the day—200,000 man-hours later—and the project was finished. It was christened ENIAC, for Electronic Numerical Integrator and Calculator.

It was an enormous, clumsy piece of machinery, weighing thirty tons and covering fifteen hundred square feet of floor space, the entire basement. Its forty panels, nine feet high, contained 18,000 vacuum tubes (the largest war radar unit used only 800), 500,000 joints soldered to connect all the circuits, in addition to 70,000 resistors and 10,000 capacitors. Three times as much electricity was needed as by a 150-kilowatt broadcasting station.

Doing away with telephone relays and other electromechanical components, Eckert and Mauchly designed their computer to perform the same task as the Harvard-IBM Mark I developed by Aiken but in a fraction of the time—about one-millionth of a second.

ENIAC was installed at the Aberdeen, Maryland, Proving Ground in 1947. It worked on problems in weather forecasting, wind tunnel design and the study of cosmic rays, as well as turning out ballistic tables for the Army and Air Force. In its initial operations it was one thousand times quicker than "Big Joel," Harvard University's wonder calculator, the most advanced general-purpose calculating machine constructed until then. The reason it would perform so fast was that it had no movable parts, a feat never before accomplished.

In one half-minute it was able to answer a problem that usually required twenty hours using a desk calculator. A five-digit number could be added five thousand times in one second. It was able to subtract, multiply, divide and extract square roots. ENIAC was so advanced that it could do the work of twenty thousand people. It solved "a very difficult wartime problem" in atomic physics in only two weeks, of which only two hours were needed on the machine, saving a year of calculations by one hundred engineers.

For the first time, a counting machine had the capacity for memory, to perform certain tasks in a proper sequence. It was remarkable to the outside world that ENIAC was able to "control" and in some ways dictate its own actions.

It received its original numbers from a series of cards in which holes were punched to indicate the "initial and boundary conditions" of a problem. When the problem was punched on the cards they were dropped into a slot in a "reader," and a unit called a "master programmer" supervised the entire computation to make sure it was carried out.

"Watch closely, you may miss it," cried one of the Moore School's faculty during an early demonstration. A button was pressed to multiply 97,367 by itself 5000 times. Most of the spectators missed it: the operation took place in less than the wink of an eye.

To illustrate ENIAC's extreme speed, the machine was slowed down by a factor of 1000 and did the same problem. It took 16⅔ minutes. The next act was to multiply 13,975 by itself. In a flash the result appeared—195,300,625. In one-tenth of a second a table of squares and cubes of numbers was produced. Next, a similar one of mathematical sines and cosines. The task was over and printed on a large sheet before most of the visitors could go from one room to another.

The ENIAC again surprised the gathering crowd when it was asked to solve a difficult problem that would have required several weeks' work by a trained man. ENIAC did it in exactly fifteen seconds.

ENIAC was built for war purposes but ironically was actually completed two months after the surrender of Japan.

Until 1946, ENIAC remained one of the country's top

secrets. It took eight years, a long period in the history of the computer, for the development of a more advanced machine to outperform ENIAC.

Both Mauchly and Eckert agreed that ENIAC need not have waited until 1945 to be constructed. This feat could have been accomplished earlier, but then there was no Second World War and a series of major military problems to solve.

In 1962, during an interview with the editor of *Datamation* magazine, Harold Bergstein asked of Eckert and Mauchly:

Since the ENIAC was a direct result of your efforts and government money during World War Two, when would you speculate that the digital computer might have been invented (a) if there had been no war, and (b) if there were no Eckert and Mauchly to invent it?

Eckert said:

I think you certainly would have had computers about the same time. There are a lot of things which cannot linger long without being born. Actually, calculus was invented simultaneously by two different individuals. It's been the history of invention over and over again when things are kind of ready for invention, then somebody does it.

What puzzles me most is that there wasn't anything in the ENIAC in the way of components that wasn't available 10 and possibly 15 years before. . . . The ENIAC could have been invented 10 or 15 years earlier and the real question is, why wasn't it done sooner?

Mauchly added:

In part, the demand wasn't there. The demand, of course, is a curious thing. People may need something without knowing that they need it.

However, Eckert and Mauchly realized that ENIAC had certain shortcomings. It could not retain a great deal of data in its memory unit; there were obstacles at the input and output points because of the high-speed operations; and the complex circuity necessitated rewriting and replugging for each new task.

A model incorporating new approaches was needed. They started to work on EDVAC, Electronic Discrete Variable Automatic Computer. They were looking for a "stored-program" to take care of the shortcomings:

The thing that made ENIAC and all the subsequent computers possible was . . . General Leslie Simon. The group that Simon assembled at Aberdeen was really a very remarkable one. Among other things he had a scientific advisory committee, the like of which I don't think we're going to see in the near future. Among others, it had a very nice young Hungarian mathematician named Johnny Von Neumann. . . .

In 1944, Goldstine met Professor Von Neumann, who came to this country in 1930 and became one of the first permanent members of the Princeton Institute for Advanced Study. One day, by chance, Goldstine saw Von Neumann at a railroad station near Aberdeen, where the Professor was also a consultant. In conversation Von Neumann learned of Mauchly and Eckert's work in building a machine one thousand times faster than any existing counter. Von Neumann found the conversation intriguing. Later Goldstine was to remark: "Once Johnny saw what we were up to, he jumped into electronic computers with both feet."

Von Neumann was an extraordinary human being. One day while at Aberdeen a "bright mathematician" was stymied by a problem and took a small computing machine home and spent half the night solving it. Von Neumann came in the next day and looked at the young man's problem, unaware that the mathematician had any solution. Attacking the problem numerically, Von Neumann threw back his head, and said:

"For n equals one," and he would talk to himself, and about one minute later he said, "17.53! Fine!" And he then said, "Now let's see what n equals two is like." After several minutes for each case, Von Neumann had knocked off n equals 2, 3, 4. Then he got to n equals 5. And this was the thing that had kept the mathematician up until 4:30, so he was really out for blood. He watched carefully as Von Neumann was going through this mumbling, and when he got to a number that this chap recognized, the fellow immediately said 67.51. Von Neumann's mouth dropped open, and he said "What?" "67.51." Von Neumann's head went back again and the calculation went much faster now. Half a minute later Von Neumann said, "67.51, that's right." This fellow ran out of the room. . . . Von Neumann was pacing back and forth in the room jiggling the keys in his pocket. You could hear him saying to himself. "How could the guy have done this?"

Dr. Von Neumann had been working on computational approaches for solving specific partial differential equations related to the atomic bomb. Convinced that a computer could be designed and constructed to be more useful to the scientist, he devised the stored-program idea of placing instructions in "memory." In discussing the task of mounting calculations, Von Neumann once said, "Probably in its execution we shall have to perform more elementary arithmetical steps than the total in all the computations performed by the human race heretofore. We noticed, however, that the total number of multiplications made by the school children of the world in the course of a few years sensibly exceeded that of our problem."

The problem with ENIAC was that each time there was a new problem, an entire set of new wires had to be introduced to enter a "program" for the needed solution. This took a considerable amount of time and effort, not to mention the chance of error. Von Neumann developed the "stored program," a system for building in certain components and basic operations, thus considerably minimizing the need for constant rewiring. The versatility of the "stored program" was spelled out by Von Neumann:

Since the orders that exercise the entire control are in the memory, a high degree of flexibility is achieved than in any previous mode of control. Indeed, the machine, under control of its orders, can extract numbers (or orders) from the memory, process them (as numbers!), and return them to the memory (to the same or other locations); i.e., it can change the contents of the memory—indeed this is its normal modus operandi. Hence it can, in particular, change the orders (since these are in the memory!)—the very orders that control its actions. Thus all sorts of sophisticated order-systems become possible, which keep successively modifying themselves and hence also the computational processes that are likewise under their control. . . . Although all of this may sound farfetched and complicated, such methods are widely used and very important in recent machine-computing—or, rather, computation-planning practice.

Von Neumann joined the project team in 1946 in collaboration with Eckert and Mauchly, the result being EDVAC.

In addition to the "stored-program" there was another major innovation. It was the use of a binary number system rather than a decimal system. Thanks to the brilliant work of Claude Shannon, a young mathematician working for the Bell Telephone Laboratories in 1937, it was now possible to demonstrate the parallel between switching circuits and the algebra of logic: true or false values were therefore analogous with open and closed states of electrical circuits. Shannon had defined a universal measure and a universal unit of information—the big (binary digit): "A bit is the choice between plus or minus; it is the amount of information needed to remove the uncertainty between yes or no."

Together, Von Neumann, Mauchly and Eckert worked in launching the EDVAC project, employing binary digits, but none remained to see it completed. A new team was brought in to take command. Von Neumann left with Goldstine to go to Princeton, where at the Institute for Advanced Study, Von Neumann continued his interest in computers and along with the RAND Corporation built the JOHNNIAC (the name was protested by Von Neumann) computer. Dr. Goldstine eventually was offered the position as director of mathematical research at the IBM Thomas J. Watson Research Center in Yorktown Heights, New York, a postition he accepted. Mauchly and Eckert left to form their own company. They took with them their years of experience at the Moore School, and in using materials similar to those found in ENIAC and EDVAC, they developed BINAC in 1948, a unit that was cheaper and faster than ENIAC and EDVAC and could handle magnetic tapes instead of punched cards. Only one BINAC was ever to be built.

Dr. Grace Hopper, a computer expert in her own right, told the story of entering Mauchly's laboratory early one morning and finding "the BINAC surrounded by Coke bottles, and sitting in front of it, slightly unshaven, John Mauchly; and both John Mauchly and BINAC singing 'Merrily We Roll Along.' "

Giant electronic brains were not particularly suited to shoestring free enterprise, and by 1951 the struggling Eckert-Mauchly Computer Corporation was a very white elephant in the hand of the Munn brothers, owners of the Totalisator Co. (racetrack tote boards). To get advice, the Munns went to their friend Jim Rand.

In 1915, after an apprenticeship with his father's company, James H. Rand, Jr., organized the American Kardex Company to manufacture his own invention, Kardex Visible Record Control System, which brought "Facts at a Glance" to the business office.

After one of the numerous arguments between Rand Senior and Junior, the elder Mrs. Rand asked, "Why don't you two boys stop fighting and instead combine your efforts and talents?" And they did. In 1925 the Rand Kardex Corporation was formed.

Young Rand, being a risk taker, went to the banks with a plan that if they would give him $25 million in cash, he would assemble the greatest office-supply company the world had ever seen and pay back the loan by selling bonds secured by the companies he bought. Jim was described as "shrewd, ruthless, persistant past the point of obstinacy."

By 1927, Remington Rand, Inc., a combination of the Remington Typewriter Company, the Rand Kardex Corporation and a half-dozen other companies, had been formed.

Remington Rand was an obvious company to go to for a merger. In 1946 it introduced the first electromechanical visible records unit, which brought automation to office records, and it established a research laboratory at South Norwalk, Connecticut, staffed by a few of the electronic engineers left over from a small wartime project on guided missiles.

By 1951 the company's laboratory had completed an electronic version of the old Powers tabulating machine.

When the Munns approached Mr. Rand he had been following the postwar work in calculator research and felt certain that sooner or later giant computers would find their way into the business field. Rand knew at that time that most computers were built by scientists for scientists to handle complex mathematics and approximations of advanced research. He was astute enough to realize, by contrast, that office management needed machines that could handle simple additions and multiplications with speed and absolute reliability. After having his engineers check into the E-M laboratories, he found that much of Eckert and Mauchly's thinking matched his own. Rand solved his friends' problem by buying E-M for a modest figure, making Eckert vice president of the computer division. But he had to pour an additional $12 million into it before a penny was realized.

Mauchly and Eckert were put in charge of the development of UNIVAC I. It was to become the first commercially feasible computer that not only could check itself for errors but could handle both numbers and descriptive data. The computer was installed in the U.S. Bureau of the Census in 1951. UNIVAC, Universal Automatic Computer, was used almost continuously twenty-four hours a day, seven days a week for more than twelve years, and was able to perform 237,000 additions of five-digit figures in one minute. In October, 1963, it was retired and placed in the Smithsonian Institution after operating more than 73,000 hours.

In 1955, James Rand's Remington Rand finally merged with Sperry Gyroscope Corporation, founded in 1910. The new corporation was to become a conglomeration of twenty-two companies. The merger began with an initial meeting of the presidents twenty years earlier, at a dance at the Westchester, New York, Sleepy Hollow Country Club. The match certainly looked like a fine one. The former was the second largest manufacturer of office equipment on a broad line and was also, at the time of the merger, number one maker of computers for business use. The latter was a leading producer of electronic weaponry, hydraulic equipment and farm machinery.

The high hopes for the Sperry Rand merger were based on two assumptions, both of which were proved to be wrong: 1) that Remington Rand as a whole was a strong organization and 2) that Remington Rand's UNIVAC operation really had a chance of maintaining its early computer lead.

In the beginning, at the time of the merger, UNIVAC was outselling IBM's 701 and 702 models. The weakness lay in Sperry Rand's inability to make, sell and maintain its computers. Meanwhile, IBM had brought out its 704 and

705 models and had begun selling these superior computers aggressively. A year after the merger, IBM was decisively ahead: in big machines it had delivered 76 to UNIVAC's 46 and had firm orders for 193 against UNIVAC's 65.

According to *Fortune* magazine's Gilbert Burck:

Few enterprises have ever turned out so excellent a product and managed it so ineptly. Univac came in too late with good models, and not at all with others; and its salesmanship and software were hardly to be mentioned in the same breath with IBM's. The upper ranks of other computer companies are studded with ex-Univac people who left in disillusionment.

Nevertheless, today one of the five hundred largest U.S. industrial corporations, Sperry Rand ranks among the top fifty in sales and assets. It employs more than 100,000 people in domestic and overseas operations. Sperry Rand products are manufactured in fifty-four plants in twenty-one states in the U.S. and forty plants in seventeen foreign countries. In the mid 1950s General Douglas MacArthur was made chairman of the board, and Jim Rand took the title of vice chairman.

Mauchly and Eckert had succeeded while others had appeared to fail. They mastered the technology; the organization to sell it to the world was faltering. At the Moore School the two men revolutionized the art of handling huge quantities of numbers in complex forms. In 1946 Eckert predicted an era in which with electronic speeds available, problems that have been thought impossible because they might require a lifetime would be readily resolved for man's use. "The old era is going," Eckert said, "the new one is electronic, speed is on the way, when we can begin all over again to tackle scientific problems with new understanding."

In 1950 Eckert and his associates solved the "Three-

Body Problem." What is the relationship between three bodies floating about in space each under its own influence of gravity?

More than eight hundred scientific reports have been completed on this subject since 1750. Until the computer was available it was an enormously lengthy project to determine all the relationships. On the basis of "good" approximations it was concluded that each of the three bodies, the sun, earth and the moon, exert little gravitational influence upon the other two bodies.

Eckert, in addition, expanded the issue to a "Six-Body Problem," to include in the relationship Jupiter, Saturn, Uranus, Neptune and Pluto—the five other major planets. With the use of computers, their positions were determined at intervals of forty days from the year 1653 and projected to the year 2000. With the computer it was now possible to feed in 25,000 observations made between 1730 and 1940, with the path of each planet calculated to the fourteenth decimal place. The final 325-page report contained 1.5 million figures and 12 million calculations made on the computer.

In a speech in 1962 before a group of industrial engineers, John Mauchly, who formed his own consulting firm, made a prediction of the future based on his present research activities. Mauchly claimed that in a decade or so everyone would own a computer with all the pertinent information needed and stored in the machine rather than his own brain. Describing a woman at a supermarket, Mauchly said:

Taking her computer from her handbag, [the housewife] enters a vacant delivery alcove and connects the computer into a receptacle provided. Within less than a minute, her packages of groceries, and other supplies such as use-once-and-throw-away clothes, come down a chute. They are assembled in advance because she had a home-data booth through which her computer, which kept her domestic inventory for her, had been able to relay in her order.

Perhaps Mauchly was a bit premature when he predicted that this event would occur about 1972, but he nevertheless was showing the way for things to come.

The world started to react to the creative innovations of Mauchly and Eckert. In comparing the UNIVAC solid-state computer with ENIAC, it is easy to see how rapidly advances were made. ENIAC, which weighed 30 tons and occupied 1500 square feet of floor space, required 18,000 vacuum tubes. The solid-state UNIVAC weighed less than 4 tons, occupied only 575 square feet of floor space and had more than 100 times the capacity of and operated 10 times faster than the original ENIAC.

In the early 1960s Mauchly and Eckert each received the John Scott Medal for "adding to the comfort, welfare and happiness of mankind."

In May, 1964, Eckert received the honorary degree of doctor of science in engineering from his alma mater, the University of Pennsylvania.

In February, 1965, the two men were corecipients of the Institute of Electrical and Electronics Engineers Philadelphia Section Award for "their fundamental concepts and contributions to electronic computers and specifically for the construction of the first all-electronic computer."

A proud moment was reached in 1966 when they both received the prestigious Harry Goode Memorial Award for "his contribution to, and pioneering efforts in, automatic computing; developing and constructing ENIAC, the world's first all-electronic computer; continuing original contributions which resulted in the development of the BINAC and UNIVAC computers, and for his continual work in the field of digital computers as an engineer."

Their place in computer history was guaranteed.

9

Alan M. Turing

The Universal Computing Machine

He was a wonderful chap in many ways. I remember how he
came to my house late one evening to talk to Professor J. Z.
Young and me after we had been to a meeting in the Philosophy
Department here, arranged by Professor Emmet. I was worried
about him because he had come hungry through the rain on his
cycle with nothing but an inadequate cape and no hat. After
midnight he went off to ride home some five miles or so through
the same winter's rain. He thought so little of the physical dis-
comfort that he did not seem to apprehend in the least degree
why we felt concerned about him, and refused all help. It was
as if he lived in a different and (I add diffidently, my impression) slightly inhuman world. Yet he had some warmth, I know
—for you in particular, for he told me so in a revealing couple
of hours that we had together not very long before he died.
. . . Alan, as I saw him, made people want to help and protect
him though he was rather insulated from human relations. Or
perhaps because of that we wanted to break through. I per-
sonally did not find him easy to get close to.

These were the carefully chosen words of Sir Geoffrey

Jefferson, written to Alan Turing's mother shortly after her son's untimely, tragic death in June, 1954.

Turing was a genius in the truest sense of the word. He was one of those one-out-of-a-thousand enormously brilliant young persons who in youth not only did things that marked him as extraordinary but possessed the ultimate awareness himself that he had performed some outstanding act.

The Turing family's history can be traced back to A.D. 1316. Of Norman descent, they first settled in Scotland. Alan's parents were of strong, upper-middle-class, well-educated, world-traveling stock. Alan Mathison Turing, their second son, was born at Warrington Crescent in London, England, on June 23, 1912.

His father was frequently away from home, spending a great deal of time during Alan's formative years in India. Around Alan's third birthday, Mrs. Turing wrote to her husband that their son was "a very clever child. I should say, with a wonderful memory for new words. Alan generally speaks correctly and well. He has rather a delightful phrase, 'for so many morrows,' which we think means 'for a long time,' and is used with reference to past or future."

His first attempt at experimentation, when he was a lad of three, was observed by his mother: "One of the wooden sailors in his toy boat had got broken, he placed the arms and legs in the garden, confident that they would grow into toy sailors."

When Alan was six years old he found a copy of *Reading Without Tears,* and taught himself the rudiments of reading in about three weeks.

At the age of eight Alan wrote what might be considered the shortest scientific book ever attempted. His manuscript, *About a Microscope,* began and ended with the sentence, "First you must see that the lite is rite."

Around this time he became preoccupied with learning

how to ride his cherished bicycle. Once he started to tackle a serious problem, he would stay with it until satisfied, or at least until he was torn away from it. One evening while he was riding around the lawn of his house darkness started to set in. Alan yelled from the distance, "I can't get off until I fall off."

His scientific curiosity began to emerge—from out of nowhere, it seemed—somewhere around his ninth birthday when he startled his mother with the inquiry: "Mother, what makes the oxygen fit so tightly into the hydrogen to produce water?"

At Hazelburst, a preparatory school, Alan tended to look upon sports as a waste of time. Years later he recalled that it was during his early school years that he learned to run fast: he was always racing to get *away* from the ball. His main joy came from studying and calculating the angle of the ball's movement. At the end of the school year his classmates joked about his interest in sports:

> Turing's fond of the football field
> For geometric problems the touch lines yield.

He had an insatiable thirst for argument. One day during a meeting of the Debating Society the motion under discussion was: "It is more interesting to keep pets than cultivate a garden." A friend argued that flowers could be taught to climb, but Turing insisted that was wrong: "Once they climb up they can't climb down." On another occasion the proposition was: "Electricity is more useful than gas." Alan said no: "Air is gas and so necessary for life."

For Christmas, 1924, his parents purchased for him a well-equipped chemistry set, and Alan began to conduct some rather sophisticated experiments in the cellar of their house. He devoted considerable time trying to extract iodine from seaweed from the local beaches. At 12½ he wrote to his mother: "I always seem to want to make things from the thing that is commonest in nature and with the least waste of energy."

But his scientific experiments were to be put aside as he began to plan for the "Common Entrance Examination." He took the tests and won entry into the Sherborne School. Alan's talents were soon recognized, and at the end of his first term he was awarded the Kirby mathematics prize for the lower school, to be followed by the Plumptre Prize for mathematics.

His mathematics master claimed that Alan was "a mathematician I think." Not yet fifteen years of age, he had evolved the calculus term "$\tan^{-1}x$," without any knowledge of calculus.

At 15½ Alan was deeply immersed in the works of the mathematicians. He wrote to his mother, trying to explain to her Einstein's books on relativity. Alan's need for clarity was brought out in his categorizing chapters, which he thought she should read:

CHAPTER XX.   "The explanation that a gravitational field is stopping the earth from moving past the train seems very arbitrary, but then people do not as a rule use railway carriages as their system."

CHAPTER XXIX.   "He [Einstein] has now got to find the general law of motion for bodies. It will have of course to satisfy the general Principle of Relativity. He does not actually give the law which I think is a pity, so I will. It is 'The separation between any two events in the history of a particle shall be a maximum or minimum when measured along its world line. . . .'"

Alan achieved the prestigious and envious position of "Mathematician-in-Ordinary" and he was in great demand for tutoring by other boys at Sherborne. In 1930 and 1931 he won the Christopher Morcom Prize for Natural Science. This award meant a great deal to Turing, for Christopher Morcom, who had died in 1930, had been his best friend at Sherborne. In a letter to his mother Alan revealed his inner feelings about Morcom:

I feel that I shall meet Morcom again, somewhere and that there will be some work for us to do together as I believed there was for us to do here. Now that I am left to do it alone I must not let him down, but put as much energy into it, if not as much interest, as if he were still here. If I succeed I shall be more fit to enjoy his company than I am now. . . . It never seems to have occurred to me to make other friends besides Morcom, he made everyone seem so ordinary.

His capturing the Morcom Prize, the Westcott House Goodman scholarship, the King Edward VI's gold medal for mathematics and acceptance into Cambridge was sufficient reason for Turing's joy as he left Sherborne.

His headmaster wrote of him:

. . . a gifted and distinguished boy . . . I have found him pleasant and friendly. . . . Turing in his sphere is one of the most distinguished boys the school has had in recent years. . . . It's a sad business, writing the last of these epistles. But there's nothing sad to look back on. Alan has done really well. A couple of years ago one doubted his capacity to be a prefect; but there's no doubt of his success and he has carried it off. Mathematicians and Scientists one is apt to regard as being soulless creatures; but Alan is not, he is warm-hearted and has a savoring humour. We shall miss him, for he was a character and won respect.

Alan was on his way to Cambridge. He had cut the last cord with his childhood, and maturity was to come to him. But although his years at Cambridge were to prove critical, the impact Sherborne had on him was unique in the school's history:

It's not however, as a great brain that I [the headmaster] shall remember him, though I acknowledge it with awe and feel a degree of humility that Westcott House should have seen him through some of his days, not without happiness. But the picture I keep and treasure is of a somewhat untidy boy arriving during the Railway Strike after making his way on a bicycle from Southampton via the best hotel in Blandford, and reporting "I am Turing." It was a good start. He was cold in various ways—as you know: as e.g., in putting butter on his porridge. But it never made him peculiar or unsocial. Neither his contemporaries, nor, I fancy, his masters knew he was of a calibre that a school is lucky to number in 100 or 200 years. But they liked him as a person and a character. I look back on holidays in Cornwall and Sark among the greatest enjoyment of my life; in all his companionship and whimsical humour, and the diffident shake of the head and rather high pitched voice as he propounded some question or objection or revealed that he had proved Euclid's postulates or was studying decadent flies—you never knew what was coming . . . what it was to be and remain human and lovable.

At Cambridge, Turing found comfort and stimulation for his exploding mind. Away from the classroom he did things in ways that startled his friends. In order to determine the correct hour when setting his watch, rather than ask a friend what time it was, he would observe a specific star, as seen from a definite place, and determine how best to adjust his timepiece.

On one occasion while playing tennis he realized that neither he nor his opponent had watches. To guarantee that they would not linger on the court too long. Turing improvised with a sundial.

But it was in the area of scientific accomplishment that Turing grew. In 1937, at the age of twenty-five years, he wrote a paper, "On Computable Numbers with an Application to the Entscheidungs-problem." The following year he produced a short work, "Computable Numbers," which was to prove to be his most famous mathematical contribution. Quoting from the London *Times:*

The discovery which will give Turing a permanent place in mathematical logic was made not long after he had graduated from Cambridge. This was his proof that there are classes of mathematical problems which cannot be solved by any fixed and definite process. The crucial step in his proof was to clarify the notion of a "definite process," which he interpreted as "something that could be done by an automatic machine." Although other proofs of insolubility were published at about the same time by other authors, the "Turing machine" has remained the most vivid, and in many ways the most convincing, interpretation of these essential equivalent theories. The description that he then gave of a "universal" computing machine was entirely theoretical in purpose, but Turing's strong interest in all kinds of practical experiment made him even then interested in the possibility of actually constructing a machine on these lines.

Alan's abstract computer, or "Turing Machine," as it was commonly referred to, represented his masterful contribution to the development of the computer. A long tape, divided into squares of 1 or 0, passed through the machine. Without any predetermined instructions the machine scanned the entire strip, one square at a time. At this point, since there were stored instructions, the machine could change a 1 to a 0, or a 0 to a 1. Likewise, it could easily move the tape in a forward or backward position by one square.

Utilizing his design, Turing was able to show mathematically that there must be universal Turing machines, or put another way, universal Turing machines that could be programmed to imitate any other Turing machine.

According to Professor Martin Davis of Yeshiva University, an authority on Turing, "the existence of universal Turing machines confirms the belief of those working with digital computers that it is possible to construct a single 'all-purpose' digital computer on which can be programmed (subject of course to limitations of time and memory capacity) any problem that could be programmed for any conceivable deterministic digital computer."

As Turing's fame spread around the world, so did the praise of his contribution to the knowledge of computers. His significance was reported years later by Dr. Robin Gandy in *Nature:*

During his first years of research he [Turing] worked on a number of subjects, including the theory of numbers and quantum mechanics, and started to build a machine for computing the Riemann Zeta-function, cutting the gears for it himself. His interest in computing led him to consider just what sort of processes could be carried out by a machine: he described a "universal" machine which, when supplied with suitable instructions, would imitate the behavior of any other; and he was thus able to give a precise definition of "computable," and to show that there are mathematical problems whose solutions are not computable in this sense. The paper which contains these results is typical of Turing's methods; starting from first principles, and using concrete illustrations, he builds up a general abstract argument.

At the age of twenty-five Alan Turing's reputation was well established. His works were interpreted and translated around the world in major journals and encyclopaedia's of the day.

Turing spent two years at Princeton University and was awarded his doctorate degree in May, 1938, the subject being "System of Logic Based on Ordinals." He was offered a post at Princeton to work as Von Neumann's assistant, but preferring the England he knew best, he returned to his home in the summer of 1938 and accepted a fellowship at King's College.

Turing was happiest with people he knew well, and in England he found himself at ease. Although he was comfortable with adults, perhaps he most found peace of mind with children. His professors' sons were among Turing's best friends. On the occasion of one of the youngster's birthdays, when asked who was coming to his party, the boy said with pride, "Six boys from the school and a bachelor—Alan Turing."

One of his neighbors had a four-year-old boy. No matter what Alan was occupied with, upon request from the lad, he would close his books and join the tot in a game or just sit and talk with him. One day the boy's mother overheard the conversation: they were debating whether God could catch a cold if He sat on the damp grass.

Throughout the World War II years, Turing served with the British Department of Communications. At the outset he was well aware that should Germany occupy England, all bank accounts would be worthless. Therefore, he converted all his savings into two ingots of silver bullion and buried each in a different location. The irony was that even with the able assistance of a homemade mine-detector, Turing was never able to relocate his bullion.

Of this and similar behavior, a friend who helped Turing search for his buried treasure said:

All these men have another thing in common which perhaps goes hand in hand with great intellect—namely, an unflagging boyish gusto over any project or topic which is raised. This was particularly so of Alan, and his enthusiasm over the treasure-hunting and chess-machine provides a good example. When I first met Alan his eccentric manner deceived me into thinking he was all head and no heart. When I knew him better I realized that his emotions were so child-like and fundamentally good as to make him a very vulnerable person in a world so largely populated by self-seekers.

From November, 1942, to March, 1943, Alan lived in the United States, and at the war's end he was made an officer of the British Empire for his participation in helping to gain victory over the enemy.

Turing was offered a Cambridge University lectureship, but this he declined. He wanted to put into practice his theory originally developed in 1937, and build his own computer.

He presented a proposal to the government and joined the staff at the National Physical Laboratory at Teddington, England. He became a permanent member of the Scientific Civil Service in October, 1945.

As a Senior Principal Scientific Officer "he threw himself into the work with enthusiasm, thoroughly enjoying the alternation of abstract questions of design with practical engineering."

By November, 1946, plans were well formulated for the construction of ACE, the Automatic Calculating Engine. According to Sir Charles Darwin, Director of the National Physical Laboratory,

the project has been picturesquely called the Electronic Brain. For a long time mathematicians have been occupied in getting better logical foundations for their subject, and in this field, about twelve years ago, a young Cambridge mathematician, by name Turing, wrote a paper which appeared in one of the mathematical journals, in which he worked out by strict logical principles how far a machine could be imagined which would imitate processes of thought.

. . . Broadly we reckon that it will be possible to do arithmetic a hundred times as fast as a human computer, and this of course means that it will be practicable to do all sorts of calculations outside the scope of human beings. The machine will have many different parts, such as circuits which do addition or multiplication, or one might give them an order like this: "Choose any number, then carry through with it a set of prescribed operations, and if the answer is bigger than seven go back and start again with a new trial number, but if it is less than seven, you are to do some other different operations." A different part of the machine is its memory, for it works so fast that there is no time to write down the answers, and therefore there must be a gear which can remember many things, so as to have them handy when they are going to be needed again for later work.

A working model, called Pilot ACE, was constructed and demonstrated publicly in the fall of 1950. In some respect this new computer was ahead of the American counterpart ENIAC. The *Times* of London on November 30, 1950, discussed Turing's computer:

The ACE itself will be built later, but the model demonstrated here to-day is none the less a complete electronic calculating machine, claimed as one of the fastest and most powerful computing devices in the world. Its function is to satisfy the ever-increasing need in science, industry, and administration, for rapid mathematical calculation which in the past, by traditional methods, would have been physically impossible or required more time than the problems justified. The speed at which this new engine works, said Dr. E. F. Bullard, F.R.S., Director of the laboratory, could perhaps be grasped from the fact that it could provide the correct answer in one minute to a problem that would occupy a mathematician for a month. In a quarter of an hour it can produce a calculation that by hand (if it were possible) would fill half a million sheets of foolscap paper.

The automatic computing engine uses pulses of electricity, generated at a rate of a million a second, to solve all calculations which resolve themselves into addition, subtraction, multiplication, and division; so that for practical purposes there is no limit to what ACE can do. On the machine the pulses are used to indicate the figure 1, while gaps represent the figure 0. All calculations are done with only these two digits in what is known as the binary scale. When a sum is put into the machine the numbers are first translated into the binary scale and coded; instructions are also given to the machine by coding them as holes in cards. To carry out long sequences of operations, the engine must be endowed with a "memory." This "memory" section is highly complicated. It depends upon the slower time of travel of supersonic waves, into which the electric pulses are converted, through a column of mercury. One thousand pulses —representing digits—can be stored in this way and extracted at the precise moment when they are needed by the "arithmetic section," which, handling pulses of electricity, is working 100,000 times faster than the supersonic section. The completed calculation appears in code as a holed card, representing the answer in the binary scale, which is translated back into ordinary numbers. When experience has been gained some improvements will doubtless be made to the Pilot ACE and embodied in the first standard prototype model. The cost of development and construction of the pilot model, which uses some 800 thermionic valves, was about £40,000. Now it is ready to "do business" and is expected to more than earn its keep.

Turing provided the theoretical work, and the engineers did the rest. ACE was unequaled for five years and became the model for future computer innovations in England. Although changes were made with time, Alan Turing was regarded as the pioneer of the computer. Borrowing an already used phrase about Charles Babbage, it was said years later, in 1958, "Today Turing's dream has come true."

Eventually, the original Pilot Ace was placed on exhibit in the London Science Museum.

Work continued on ACE, but progress was too slow for Turing. It was during this period that he decided to get away from ACE and took a year's sabbatical at King's College.

Around 1944, Turing began to theorize about the construction of a universal computer and of the service such a machine could perform for psychology in explaining human behavior and the performance of the brain.

He shared his experiences and ideas with Professor Norbert Weiner of MIT, who defined cybernetics as the science of "control and communication in the animal and the machine."

Turing began to draw parallels between the operation

of the brain and that of a computer. He wrote a paper, "Can a Machine Think?" which was discussed in *The World of Mathematics,* whose editor wrote:

Can machines think? Is the question itself, for that matter, more than a journalist's gambit? The English logician, A. M. Turing, regards it as a serious, meaningful question and one which can now be answered. He thinks that machines can think. He suggests that they can learn, that they can be built so as to be able to do more than we know how to order them to do, that they may eventually "compete with men in all purely intellectual fields."

Alan Turing had thought out the answer to the question "Can a machine think?" and was well prepared to answer his critics. To the doubters he said:

Whenever one of these machines is asked the appropriate critical question and gives a definite answer, we know that this answer must be wrong, and this gives us a certain feeling of superiority. Is this feeling illusory? It is no doubt quite genuine, but I do not think too much importance should be attached to it. We too often give wrong answers to questions ourselves to be justified in being very pleased at such evidence of fallibility on the part of the machines. Further, our superiority can only be felt on such an occasion in relation to the one machine over which we have scored our petty triumph. There would be no question of triumphing simultaneously over all machines. In short then, there might be men cleverer than any given machine, but, then again, there might be other machines, cleverer again, and so on.

Turing set out to prove to the world that he was right. Starting with 1949, he was appointed assistant director of the Manchester Automatic Digital Machine (MADAM), reputed to have the largest memory storage capacity of any known machine at that time.

Alan Turing was well ahead of his day, and his prophecies about computers appeared to be proving right as each new year brought further advances in hardware sophistication. In a brilliant essay, Turing had some well chosen words for us:

We may hope that machines will eventually compete with men in all purely intellectual fields. But which are the best ones to start with? Even this is a difficult decision. Many people think that a very abstract activity, like the playing of chess, would be best. It can also be maintained that it is best to provide the machine with the best sense organs that money can buy, and then teach it to understand and speak English. This process could follow the normal teaching of a child. Things could be pointed out and named, etc. Again I do not know what the right answer is, but I think both approaches should be tried. We can only see a short distance ahead, but we can see plenty there that needs to be done.

One of his closest friends, the English mathematician M. H. A. Newman, in reflecting about Alan's genius, said:

I remember sitting in our garden at Bowdon about 1949 while Alan and my husband discussed the machine [the Manchester Automatic Digital Machine] and its futuristic activities. I couldn't take part in the discussion and it was one of many that had passed over my head, but suddenly my ear picked up a remark which sent a shiver down my back. Alan said reflectively, "I suppose when it gets to that stage we shan't know how it does it."

Throughout the mid and late 1940s, scientists throughout the world applied Turing's concept of mathematics and computers. They were particularly useful to the cybernetics team of Dr. Warren McCulloch, a psychiatrist from the University of Illinois, and Walter Pitts, a mathematical logician at Chicago University. They believed that any functioning that can be defined at all logically, strictly and without ambiguity in a definite number of words could be identified in a fixed system.

The echo of Turing was applied in the neural-physiological field, and the highly respected McCulloch-Pitts Theory of Formal Neural Network evolved.

In 1947 they designed equipment to enable the blind to "read" the printed page by ear and built a theory of vision. Advancing Turing's model, they were able to show that the nervous system functions in a way similar to mathematical logic.

At the turn of the decade, in 1951 and 1952, Turing made a series of radio broadcasts on computers and their ability to think. At the end of one of the debates, Professor Geoffrey Jefferson turned and said: "It would be fun some day, Turing, to listen to a discussion, say on the Fourth Programme, between two machines on why human beings think that they think."

Although Sir Geoffrey disagreed with Alan, there was no doubt in his mind that Turing was among the giants in the world: "Alan in whom the lamp of genius burned so bright . . . he had real genius, it shone from him."

Turing and Dr. Robin Gandy spent a weekend preparing a nonpoisonous weed killer and sink cleaner. Alan got a "kick" out of the play work and liked the idea of making his own chemicals rather than purchasing them.

Turing was also engaged in developing some electrolysis experiment. He was plating spoons with potassium cyanide and other chemicals. "When we worked together," said Gandy "on some electrical contraption he several times got high voltage shocks by sheer carelessness."

On June 8, 1954, Turing was found dead in bed. The death was caused by poisoning from potassium cyanide. The verdict at the inquest said that the poison was self-administered.

Perhaps it will never be known why Turing died. He was at his peak; he had attained worldwide reputation, sufficient monies and the potential for still further achievement.

Many of his friends, including his mother, would not accept the verdict of suicide. For them his death was caused by some unaccountable venture.

His mother established as a permanent memorial the Alan Turing Prize for Science awarded annually at the Sherborne School. In honor of Alan Turing the Sherborne School named their new science building The Alan Turing Laboratories.

But Turing himself will be shared by the entire industrial world. He was a pioneering giant in the computer field and made a permanent contribution to mathematical understanding by his logical analysis of the computing process.

He had set out a "paper machine," as he called it, which would be a set of instructions and approaches that could be carried out by someone, completely ignorant of the purpose and procedure involved. In order to show us what machines were capable of doing, he strived to demonstrate that a universal machine could exist that was able to perform any computation that any other machine could perform.

He devoted a major portion of his life trying to answer the question, Can the machine think? He succeeded in proving that machines can perform deductive analysis by solving mathematical equations and making logical decisions.

Only time, the future and the work of others will show how close Turing came to uncovering some of the hidden secrets of man's genius.

In describing her son's story, Mrs. Turing concludes with a brief quotation:

> He does not sleep,
> How could that eager mind be stilled by death?

10

Robert M. Fano

The Coming of MAC

"If we really want to see the computer serve mankind, we've got to find a way for everyone to use it," said an officer of the U.S. Department of Defense. "We've got to make computers directly useful to more and more people."

The intellectual and technical community, and the country's defense experts, were convinced that in the last several years the purpose of computers had not kept up with their promise. Computer leaders were frustrated with the difficulty that prevented them from attaining a closer relationship with computers. There is a very real and discouraging "communication barrier" between man and machine.

Up to this time a highly trained programmer was needed to operate the computer. Would the day come when the head of a household, possessing his own small computer, would be able to balance his bank book, pay his taxes, utility, doctor and telephone bills without ever leaving his house?

The goal, according to members of the Advanced Research Project Agency of the U.S. Government, could be reached where people would be able to handle large masses of data with no more effort than that of writing words on a typewriter.

In 1961, Professor John McCarthy of the MIT faculty presented a lecture that summarized the feelings of some of his colleagues about the future computational needs of the institute. McCarthy called for machine-aided cognition in computers—meaning the close relationship between a person using the computer in a real-time instantaneous dialogue.

The ideas were developed by the MIT Computation Center headed by Electrical Engineer Professor F. J. Corbato. CTSS—Compatible Time-sharing System—resulted and was first demonstrated in November, 1961.

Under a $2.2 million, eighteen-month contract, the Massachusetts Institute of Technology initiated Project MAC—a double-barreled acronym standing for Machine Aided Cognition, which was the objective of the project, and Multiple Access Computer, which was the principal tool employed. MAC was therefore to be the direct descendant of CTSS and was initiated in the spring of 1963.

Dr. Robert Fano, Ford Professor of Engineering and professor of electric communications and head of the MIT program, was well aware of the present major drawback in the value of computers: computers "communicate" in numbers while people speak words.

Artificial languages had overcome some of the barriers but had only been of little assistance. It still required several days or weeks to prepare an appropriate program involving a tediously slow process, the end result being that only a few people could use a computer at any one given time. It was virtually impossible for a scientist to "ask a simple question and get a simple answer."

"If the purpose of Project MAC can be summed up in a

sentence, it is to remove the barriers and give the machine to the people." This meant perfecting the principle of time sharing.

Professor Fano looked at the project in terms of "thinking of the computer as akin to the public utility, a major service whose capabilities can be made available to many people."

Dr. Fernando J. Corbató was given the responsibility for the development of the system, and Dr. Fano was to head the over-all MAC project.

Robert M. Fano was born on November 11, 1917, in Torino, Italy. His father, Gino Fano, was at one time professor of mathematics at the University of Torino.

After attending the highly respected School of Engineering of Torino, Robert came to the United States in 1939. In 1941 he received his bachelor of science degree from MIT and in 1947 his doctorate of science in electrical engineering. His rise to the top of the institute's ranks was quick. He started as an assistant in electrical engineering in 1941, then became an instructor in 1943, a research associate in 1945, an assistant professor of electrical communications in 1947, associate professor in 1951 and professor in 1956. In May, 1962, he was appointed Ford Professor of Engineering.

During the Second World War he worked at MIT's radiation laboratory on microwave components and filters, and upon its establishment in 1946 was a staff member of the research laboratory of electronics.

Married and the father of three children, he lives in the historic Lexington, Massachusetts, not far from his home base at MIT.

Bob Fano was well prepared for Project MAC. He had already developed a worldwide reputation and was known as a man of high integrity, dedication and social understanding. Meeting upon meeting was held to plan the strategy of MAC. Brainstorm session followed brainstorm session. Before, a computer user prepared in careful detail a program covering the total range of the problem he wanted to solve. Then he had to wait his turn by signing up for time on the computer.

The conventional system, known as "batch processing," required the scientist to submit his problem, usually in the form of punched cards, to a member of the computer staff, who then gathered the problems in batches, transferred them to a magnetic tape and fed them through the computer sequentially.

"Computers are like gold mines," said Dr. Licklinder, the Advanced Research Project Agency project director, "and we're getting relatively little out of them. Scientists have to line up outside and go in to pick at the gold one at a time."

The success of Project MAC demanded solving through two problems—developing easy and simple techniques for communicating with computers, and finding multiple uses of the computer, to share a computer's time.

"If this research program is successful," said Dr. Licklinder, "the implications may be far-reaching, extending beyond defense into business, education, research, engineering, and indeed almost every activity involving men and information."

According to Fano and his team, there were other ways to utilize the time of a computer. Electronically it could store in its memory more than one program and could tackle two or more of them at the same time. In a more sophisticated way there was no reason why a computer couldn't work on one program for a fraction of a second, or several seconds, then transfer to another task.

Project MAC succeeded, while others failed. Using forty different teleprinter consoles placed throughout the campus, staff members were able to communicate with either of the two IBM 7094 computers. Contacts were made through the institute's telephone system by dialing an appropriate extension number.

The heart of the time-sharing system is another small computer hooked up to the larger one and acting as a sort of "receptionist." According to Fano, the smaller computer "listens" to all the requests for data, keeps a tab of who's been answered what and lets each applicant go in to see the "boss" in his turn. While the small computer is working, the big one is carrying out its computation in four-millionths of a second. Both computers had 64,000-word memory units, double that of previous units.

After forty terminals the plan was to permit one hundred terminal hook-ups to the computers. "This is a way of bringing the larger power of computers to the people who need it," Professor Fano explained, "where they need it and when they need it."

MAC started to yield results in the fall of 1963, just one week after the largest IBM 7094 computer at that time had been installed. The end result: as many as forty different people or teams could utilize the computer at the same time with only the shortest of delays (which resulted from the computer's changing attention from one person to another).

The more complex problem was communicating with the computer, not in an artificial language but in plain English. In mid-January of 1964 an experimental demonstration was given in which the computer was asked to solve a very simple problem. It was given—in ordinary sentences—the gas consumption of a car and the distance between two cities, and was asked for the amount of gas needed for the trip.

Working twenty-four hours a day, seven days a week, MAC operated by rotating its attention to all the stations within the network. After identifying the orders, it "thinks through" the problem in one-twentieth of a second, then proceeds to store the partial solution. Next it moves to the following station for one-twentieth of a second, and so on. When a problem is completed the solution is printed out on the teleprinter at the rate of one hundred words per minute.

Within the project there were five levels of priority and therefore few problems took longer than ten seconds to be resolved with the computer.

So selective is the operation that when an unusual or rare problem demanding billions of calculations—a problem that would occupy the machine's time all day—the task is pushed in the "background," to be worked upon when the computer has the spare time.

Dr. Fano predicted that within a few years "one can type simple English instruction on the teleprinter and receive simple English replies."

In predicting the future of technology, Professor Fano contrasted the computer of the day to the interregnum between the early steam engine and the electric power utility: "Steam engines made possible fast public transportation and central power stations. But when electricity was distributed to homes, offices and factories, it made this power directly available to the individual." The goal of Project MAC's team was to make this computer power available to anyone interested in using it. The day was predicted where computer consoles would be as common as are home telephones.

As one of the columnists of the day said about Project MAC, ". . . man is still trying to 'pick the brain' that he himself created."

J. Weizenbaum, working with Project MAC, developed a conversation carried out with a program.

WHICH SCRIPT PLEASE?
Arithm.

PLEASE BEGIN.
The area of a sphere is pi times the diameter of the sphere squared.
The area of a circle is pi times the radius of the circle squared.
The diameter of an object is twice the radius of the object.
The radius of a ball is 10.
What is the area of the ball?

THE ASSUMPTION THAT BALL IS SPHERE AND OBJECT IS CONSISTENT—SHALL I GO ON?
Yes.
IT'S 1256.6359.
What is the diameter of the ball?
IT'S 20.
The radius of a dish is 10.
A dish is an object.
What is the diameter of the dish?
IT'S 20.
What is the area of the dish?

THE ASSUMPTION THAT DISH IS SPHERE IS CONSISTENT—SHALL I GO ON?
No.

THE ASSUMPTION THAT DISH IS CIRCLE IS CONSISTENT—SHALL I GO ON?
Yes.
IT'S 314.15899.
The radius of a globe is 10.
A globe is a sphere.
What is the area of the globe?

THE ASSUMPTION THAT GLOBE IS OBJECT IS CONSISTENT—SHALL I GO ON?
Yes.
IT'S 1256.6359.
The radius of a thing is 10.
A sphere is an object.
A thing is a sphere.

What is the area of the thing?
IT'S 1256.6359.
Thank you.
YOU'RE QUITE WELCOME, COME AGAIN.

Fano had a very specific objective. "What we are trying to do is bring computers into the lines of creative people, so that using the intellectual amplification that a computer provides will be as easy as reaching for a typewriter or telephone." He and his team fulfilled their expectation.

In an article they coauthored, Fano and E. E. David, Jr., of the Bell Telephone Laboratories, said the following about accessible computing:

. . . such service will provide to the individual "thinking tools," somewhat analogous to power tools, to aid him in his daily intellectual labor. These thinking tools will increase the power, skill and precision of his mind, just as power tools today increase the power, precision and skill of his muscles. As a matter of fact, there is some question whether our increasingly complex society can survive much longer without falling apart from its own weight, unless individual thinking aids become available. At the same time, the benefits they may bring to society will unquestionably be mixed with a dose of new problems and frustrations.

As sensitive social oriented scientists Fano and David were aware that the work of MAC would enhance the computer's ability to collect and store more and more information about people:

The very power of advanced computer systems makes them a serious threat to the privacy of the individual. If every significant action is recorded in the mass memory of a community computer system, and programs are available for analyzing them, the daily activities of each individual could become open to scrutiny.
While the technical means may be available for preventing

illegal searches, where will society draw the line between legal and illegal? Will the custodians of the system be able to resist pressure from government agencies, special interest groups and powerful individuals? And what about the custodians themselves? Can society trust them with so much power?

Philosophically, Fano and David saw nothing in the use of computers that would impersonalize or mechanize individual behavior. "The danger lies in ourselves. Through mental laziness, or fear of accepting responsibility, or just plain neglect, we may delegate to computers prerogatives that should remain ours."

Dr. Fano was not a genius only for the moment but could look around the corner and see applications of his work. He could with ease predict the future and counteract those who would discount the value and resist the technology:

One can think of many other instances in our society where accessible computing service, with the appropriate software, could help individuals to contend more successfully and with less frustration with the complexities of the modern world: from paying bills and balancing one's bank account to planning a will; from budgeting the family income to selecting investments and making plans for retirement. It may seem strange at this time to envision the average man and housewife using a computer. Yet, to some people years ago it must have seemed equally inconceivable and perhaps sacrilegious to allow the average housewife to turn on powerful motors and operate such complex machines as today's automatic washing machines and driers. Not many years ago we would have winced at the thought of allowing teen-agers to spend hours monopolizing such a priceless creation of human inventiveness and technology as the telephone.

Robert Fano and his associates have made that important inroad of successfully handling enormous information, at a low cost, available to a large audience. From the days of Pascal to the present, a severe limitation with calculations has been the slowness of the operation. Data has the unfortunate characteristic of usually becoming obsolete and outdated, difficult to pin down or locate, and time consuming to retrieve. These computer pioneers have devoted their lives to resolving this burdensome problem. They have not only succeeded, in their own way, to make information more accessible, but in their endeavors they have shown us new paths for improving one of the least-understood phenomena facing man—communications.

With increased data and the tool for mass communication, the day may not be far off when we reexamine why we have come to praise these men, not only for their inventiveness and enthusiasm, but for the benefit they have given us in trying to elevate mankind to a higher, more responsible level of humanness. When we all turn to the computer the same way we do a television set or automobile, perhaps then we will have understood what the quest has been for all the computer prophets.

Suggested Reading

For the reader who wishes to learn more about computers and the people responsible for their development, a list of references appears below:

I. Computers

Adler, I., *Thinking Machines,* New York: John Day, 1961, 189pp.

Andrew, A. M., *Brains and Computers,* Toronto: Harrap, 1963, 78pp.

Arden, B. W., *An Introduction to Digital Computing,* Reading, Mass.: Addison-Wesley, 1963, 389pp.

Arnold, Robert, Harold C. Hill, Aylmer V. Nichols, *Introduction to Data Processing,* New York: John Wiley and Sons, 1966, 326pp.

Berkeley, Edmund C., *Giant Brains: or Machines That Think,* New York: John Wiley and Sons, 1949, 294pp.

Bernstein, Jeremy, *The Analytical Engine,* New York: Random House, 1963, 113pp.

Crowley, Thomas H., *Understanding Computers,* New York: McGraw-Hill, 1967, 142pp.

Davidson, C. H. and Eldo C. Koenig, *Computers: Introduction to Computers and Applied Computing Concepts,* New York: John Wiley and Sons, 1967, 596pp.

Desmonde, W. H., *Computers and Their Uses,* Englewood Cliffs, New Jersey: Prentice-Hall, 1964, 296pp.

Fink, Donald G., *Computers and the Human Mind: An Introduction to Artificial Intelligence,* Science Study Series, Garden City, New York: Doubleday Anchor Books, 1966, 301pp.

Haga, Enoch, *Understanding Automation,* Elmhurst, Ill.: The Business Press, 1965, 437pp.

Hollingdale, S. H. and G. C. Tootill, *Electronic Computers,* Middlesex, England: Pelican, 1965, 336pp.

Pfeiffer, J. P., *The Thinking Machine,* Philadelphia: Lippincott, 1962, 242pp.

Postley, John A., *Computers and People,* New York: McGraw-Hill, 1960, 246pp.

Van Ness, Robert G., *Principles of Punched Card Data Processing,* Elmhurst, Ill.: The Business Press, 1962, 263pp.

II. People in History

"A History of Sperry Rand Corporation," *Sperry Rand,* Recording and Statistical Division of Sperry Rand Corporation, Dec. 1962, pp. 2–32.

Babbage, Charles, *Charles Babbage and His Calculating Engines—Selected Writings,* edited by Philip and Emily Morrison, New York: Dover Publications, Inc., 1961, 400pp.

Belden, Thomas and Marva, *The Life of Thomas J. Watson,* Boston: Little, Brown and Co., 1962, 332pp.

Caillet, Emile, *Pascal—The Emergence of Genius,* New York: Harper and Row, 1961, 383pp.

Derry, T. K., and T. I. Williams, *A Short History of Technology,* Cambridge: Oxford University Press, 1960.

Fano, Robert and F. J. Corbató, "Time—Sharing on Computers," *Scientific American,* September 1966.

Felt, D. E., "Mechanical Arithmetic or The History of the Counting Machine," Lectures on Business.

Grant, George, "Calculating Machine," *Journal of the Franklin Institute,* Vol. 60, 1855, p. 391.

Grant, George, "Calculating Machine," *American Journal of Science,* Vol. 108, 1874, p. 277.

Johnson, E., *A Brief Account of the Life, Writings and Inventions of Sir Samuel Morland,* Cambridge: Trinity Street, and London: Whittaker and Co., 1838, 31pp.

Klemm, Friedrich, *History of Western Technology,* Translated by Dorothea W. Singer, New York: Scribner, 1959.

Maboth, Moseley, *Irascible Genius, A Life of Charles Babbage, Inventor,* Hutchinson of London, 1964, p. 288.

Mantoux, P. *The Industrial Revolution in the 18th Century,* Macmillan, 1961.

Merz, John T., *Leibniz,* Edinburgh and London: Wm. Blackwood and Sons, 1884, 216pp.

Mesnard, Jean, *Pascal—His Life and Work,* London: Harvill Press, 1952, 211pp.

Morison, Elting E., *Men, Machines and Modern Times,* Cambridge, Mass.: The M.I.T. Press, 1966.

Russell, Bertrand, *A Critical Exposition of the Philosophy of Leibniz,* London: Allen and Unwin, 1951, 311pp.

Singer, C. J., et al., editors, *History of Technology,* 5 volumes, Clarendon Press, 1954–60.

Soulard, R., *A History of the Machine,* New York: Hawthorn, 1963.

Turing, Sara, *Alan M. Turing,* Cambridge: W. Heffer and Sons, 1959.

Watson, Thomas J., Jr., *A Business and Its Belief—The Ideas that Helped Build IBM,* New York: McGraw-Hill Book Co., 1963, 107pp.